Understanding Parallel
Supercomputing

IEEE PRESS UNDERSTANDING SCIENCE & TECHNOLOGY SERIES

The IEEE Press Understanding Science & Technology Series treats important topics in science and technology in a simple and easy-to-understand manner. Designed expressly for the nonspecialist engineer, scientist, or technician, as well as the technologically curious—each volume stresses practical information over mathematical theorems and complicated derivations.

Books in the Series

Blyler, J. and Ray, G., *What's Size Got To Do With It? Understanding Computer Rightsizing*

Deutsch, S., *Understanding the Nervous System: An Engineering Perspective*

Evans, B., *Understanding Digital TV: The Route to HDTV*

Gregg, J., *Ones and Zeros: Understanding Boolean Algebra, Digital Circuits, and the Logic of Sets*

Hecht, J., *Understanding Lasers: An Entry-Level Guide,* 2nd Edition

Hord, R. M., *Understanding Parallel Supercomputing*

Kamm, L., *Understanding Electro-Mechanical Engineering: An Introduction to Mechatronics*

Kartalopoulos, S. V., *Understanding Neural Networks and Fuzzy Logic: Basic Concepts and Applications*

Lebow, I., *Understanding Digital Transmission and Recording*

Nellist, J. G., *Understanding Telecommunications and Lightwave Systems,* 2nd Edition

Sigfried, S., *Understanding Object-Oriented Software Engineering*

Ideas for future topics and authorship inquiries are welcome. Please write to IEEE PRESS: The Understanding Series.

Understanding Parallel Supercomputing

R. Michael Hord

IEEE Press Understanding Science & Technology Series
Dr. Mohamed E. El-Hawary, *Series Editor*

The Institute of Electrical and Electronics Engineers, Inc., New York

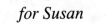

for Susan

This book may be purchased at a discount from the publisher
when ordered in bulk quantities. Contact:

IEEE Press Marketing
Attn: Special Sales
445 Hoes Lane, P.O. Box 1331
Piscataway, NJ 08855-1331
Fax: 1-732-981-9334

For more information on the IEEE Press,
visit the IEEE home page: http://www.ieee.org/

Printed in the United States of America
10   9   8   7   6   5   4   3   2   1

ISBN 0-7803-1120-5

IEEE Order Number: PP4721

Library of Congress Cataloging-in-Publication Data

Hord, R. Michael, 1940–
    Understanding parallel supercomputing / R. Michael Hord.
        p.   cm. — (IEEE Press understanding science & technology
series)
    Includes bibliographical references and index.
    ISBN 0-7803-1120-5 (paper)
    1. Parallel processing (Electronic computers)   2. Supercomputers.
I. Title. II. Series.
QA76.58.H68   1998
004'.35—dc21
                                                98-6334
                                                  CIP

Contents

Foreword

We live in a period of computer architecture innovation. The advent of parallel processing has produced a flood of architecture paradigms, some more successful than others.

The most well-known taxonomy for parallel computers was proposed by M.J. Flynn in 1966. It is based on the multiplicity of the instruction and data streams. Hence, this taxonomy identifies four classes of computers: (1) single instruction stream single data stream (SISD) computers correspond to regular von Neumann nonparallel computers that can execute one instruction at a time on one data item at a time; (2) single instruction multiple data (SIMD) computers are based on a central program controller that drives the program flow and a set of processing elements that all execute the instructions from the central controller on their individual data items; (3) multiple instruction single data (MISD) computers pass intermediate results to other processors for further action so it is freed to perform its operations on the next item in the data stream; and (4) multiple instruction multiple data (MIMD) computers consist of a number of computers configured to communicate among themselves in the course of a program, with each executing its own processors.

Examples of successful SIMD computers are too numerous to list. Instances that are discussed in this book include the Illiac IV (I4), the Massively Parallel Processor (MPP), the GAPP, the Distributed Array Processor (DAP), the MasPar (MP-2), the ASPRO, the Connection Machine-2 (CM-2); other examples include the Wavetracer and the Princeton Engine.

In this book, we also explore the MIMD category but, of course, this category is not monolithic; it exhibits great diversity, and we can only examine the high-performance subset of this rich and thriving group of architectures. It is only a snapshot since new computing paradigms are conceived and produced with great frequency, a phenomenon that can only be explained by the handsome economic rewards that accrue to anyone who can devise a way to achieve more instructions per dollar.

Using parallel machines effectively can be a considerable challenge. Just as instructing a single bricklayer exercises less managerial skill than being the foreman of a crew constructing a high-rise building, programming a parallel supercomputer requires a great deal more planning, coordination, and talent than developing an application on a sequential machine. In time, practitioners acquire the ability to "think parallel"; however, to remain competitive, programmers need this skill more and more because the movement toward parallel processing is inexorable.

Much of the material included in this book is based on or adopted from work previously published in government reports, journal articles, and vendor literature. A listing of the major sources is provided in the text.

R. Michael Hord

Preface

Understanding Parallel Supercomputing is a survey book providing a thorough introductory review of high-performance parallel computers, a type of computer that has grown to importance in recent years. It was written to describe a technology in-depth including the architectural concepts, a variety of hardware implementations, (both commercial and research), major programming concepts, algorithmic methods, and representative applications.

The book is intended for a wide range of readers. Computer professionals will find sufficient detail to incorporate much of this material into their own endeavors. Program managers and applications system designers may find the solution to their requirements for high-computational performance at an affordable cost. Scientists and engineers will find sufficient processing speed to make interactive simulation a practical adjunct to theory and experiment. Entrepreneurs will find a case study of an emerging and maturing technology. The general reader is afforded an opportunity to appreciate the power of advanced computing and some of the ramifications of this growing capability.

Although there are numerous books on conventional parallel processing, this volume is devoted to supercomputing on a wide variety of parallel machines. The reader already familiar with sequential processing will discover an alternative philosophy—parallelism—including both the widespread program parallel paradigm and the data parallel scheme.

The contents of the book are organized into nine chapters, rich with illustrations and tables. Topics covered include: High-Performance Computing and Communications Program, Grand Challenges, Parallel Supercomputing Concepts, History, SIMD and MIMD (e.g., Paragon, T3D, CM-5, etc.), Alternative Architectures, Distributed Heterogeneous Supercomputing, Software, and Applications.

Chapter 1

Introduction

As demands emerge for ever greater computer processing speeds and capacities, traditional serial processors have begun to encounter physical laws that inhibit further speed increases. One impediment is the speed of light. Signals cannot propagate faster than about a foot in a nanosecond. Hence, computer components commanded by another component ten feet away cannot respond in less than 10 ns. To defeat this limit, designers try to make computers smaller. This effort encounters limits on the allowed smallness of chip feature sizes and the need to dissipate heat.

The most promising strategy to date for overcoming these limits is the abandonment of serial processing in favor of parallel processing. Parallel processing is the use of multiple processors simultaneously working on one problem. The hope is that if a single processor can generate X floating point operations per second (FLOPS), then ten of these may be able to produce 10X FLOPS, and, in the case of massively parallel processing, a thousand processors may produce 1000X FLOPS.

Problems of obvious interest for parallel processing because of their computational intensity include the following.

- Matrix inversion
- Artificial vision
- Data base searches
- Finite element analysis
- Computational fluid dynamics
- Simulation
- Optical ray trace
- Signal processing
- Optimization

However, the range of applications for parallel processing has proven to be much broader than expected. This volume examines various types of parallel processing and describes by example the wide variety of application areas that have shown themselves to be well-suited to parallel processing.

Parallel processing, or concurrent computing as it is sometimes termed, is not conceptually new. For as long as there have been jobs that can be broken into multiple tasks that in turn can be handed out to individual workers for simultaneous performance, team projects have been an effective way to achieve schedule speedup. In the realm of computation, one recalls the depression era U.S. government's Works Progress Administration (WPA) projects of the 1930s to generate trigonometric and logarithmic tables that employed hundreds of mathematicians, each calculating a small portion of the total work. Lenses were designed the same way, with each optical engineer tracking one ray through a candidate design.

The recent excitement for parallel computer architectures results from the rising demand for supercomputer performance and the simultaneous maturing of constituent computer technologies that make parallel processing supercomputers a viable possibility.

The term supercomputer enjoys an evolving definition. It has been facetiously defined as those computers that exhibit throughput rates 50% greater than the highest rate currently available. The advent of the term occurred in the 1975 time frame, when it was variously applied to the CDC-7600, the Illiac IV, and other high-performance machines of the day. The usage became firmly established on the arrival of the Cray-1. Today, with the need for high performance computing greater than ever, the supercomputer identifier is commonplace. For the purpose of this book the term supercomputer means that class of computers that share the features of high speed and large capacity compared with the average of what is available at any given time. Both elements are important: High speed on small problems is insufficient. With this definition supercomputers over time cease being supercomputers and retire to the category "former supercomputers."

Today there are hundreds of supercomputers in the world because there are many applications that offer a very high payoff despite the large costs of supercomputing methods. For example, consider the case described by W. Ballhaus of the NASA Ames Research Center (Mountainview, CA): Supercomputers were used to design the A-310 Airbus Airliner. It is estimated that the use of the supercomputer resulted in a fuel efficiency improvement of 20%. For jet fuel priced at $1.30/gal, over a 15-year life expectancy with 400 flights per year per plane, the fleet life fuel savings resulting from the improved efficiency is $10 billion.

Parallel processing supercomputers have not always been technically feasible. They require interprocessor communication to perform sufficiently well that multiple processors can execute an application more quickly than a single processor acting alone can execute that application. Even today we see cases where 32 processors are slower than 16 processors working the same problem, not because the problem is insufficiently parallel, but because the interprocessor communications overhead is too high. These cases are becoming less common as the various constituent technologies mature.

Another technology maturity issue making parallel processing supercomputing feasible today is that the cost of implementation is more and more affordable. VLSI chip technology is revolutionizing the cost-performance characteristics of recent systems.

Parallel computers have been evolving for 30 years. Today they have become an essential and undeniable force in large-scale computing. This book explores their design, their programming methods, and a selection of their applications in some depth.

We begin by examining the parallel supercomputing environment on a large scale. Specifically, we consider two areas, the role the U.S. government has taken to foster the high performance parallel computer technology; and the grand challenge problems, that is, a list of answers to the questions "For what uses are we developing these powerful computational engines? Why do we need this much computing power?"

The role of the U.S. government has been and continues to be key to the existence and growth of parallel supercomputing. Multibillion dollar investments have led the technology to its present state. Previously the U.S. government's performance objective was a thousand billion floating point operations per second by a single multiprocessor computer; plans in 1998 are being discussed for the following hardware/software generation's objective of a million billion floating point operations per second.

The grand challenge problems are described to explain the investments, by the U.S. government, industry, and academia, in parallel supercomputing. It is by no means the case that one needs a grand challenge problem to justify one's use of a parallel supercomputer, as the later chapter on applications will demonstrate; or that the list of grand challenge problems described is an exhaustive enumeration of grand challenges.

Together these expositions set the stage for the technical chapters that follow.

1.1   HIGH PERFORMANCE COMPUTING AND COMMUNICATIONS PROGRAM

The goal of the Federal High Performance Computing and Communications (HPCC) Program is to accelerate significantly the availability and utilization of the next generation of high performance computers and networks in a manner consistent with the Strategic and Integrating Priorities shown in Figure 1.1-1.

The HPCC Program is the result of several years of effort on the part of senior government, industry, and academic scientists and managers to design a research agenda to extend U.S. leadership in high performance computing and networking technologies. The program is managed with close cooperation among federal

Goals: Strategic Priorities
• Extend U.S. technological leadership in high performance computing and computer communications. • Provide wide dissemination and application of the technologies both to speed the pace of innovation and to serve the national economy, national security, education, and the global environment. • Spur gains in U.S. productivity and industrial competitiveness by making high performance computing and networking technologies an integral part of the design and production process.
Strategy: Integrating Priorities
• Support solutions to important scientific and technical challenges through a vigorous Research and Design effort. • Reduce the uncertainties to industry for Research and Design and use of this technology through increased cooperation among government, industry, and universities and by the continued use of government and government-funded facilities as a prototype user for early commercial HPCC products. • Support the underlying research, network, and computational infrastructures on which U.S. high performance computing technology is based. • Support the U.S. human resource base to meet the needs of industry, universities, and government.
Program Components

High Performance Computing Systems (HPCS)
    Research for Future Generations of Computing Systems
    System Design Tools
    Advanced Prototype Systems
    Evaluation of Early Systems
Advanced Software Technology and Algorithms (ASTA)
    Software Support for Grand Challenges
    Software Components and Tools
    Computational Techniques
    High Performance Computing Research Centers
National Research and Education Network (NREN)
    Interagency Interim NREN
    Gigabits Research and Development
Basic Research and Human Resources (BRHR)
    Basic Research
    Research Participation and Training
    Infrastructure
    Education, Training, and Curriculum

Figure 1.1-1.   The High Performance Computing and Communications Program

agencies and laboratories, private industry, and academe to ensure that the fruits of this research program are brought into the educational and commercial marketplaces as rapidly as possible.

There have been steady accomplishments over the past. Several participating agencies fund HPCC research groups, centers, and consortia on various grand challenge application problems. Major new scalable high performance systems have been announced and delivered. New software applications have been developed or ported to the emerging high performance systems. Traffic on the operational network has grown rapidly, as has the number of interconnected local and regional networks. A large number of researchers, scholars, students, educators, scientists, and engineers have been trained to use these emerging technologies.

Numerous organizations have undertaken studies that have provided valuable feedback on the structure and content of the program. Some of the organizations that supported studies include the National Academy of Sciences; EDUCOM and the Computing Research Association, each representing numerous universities; the Computer Systems Policy Project, representing leading U.S. computer companies; the Office of Technology Assessment; professional societies including Association for Computing Machinery, Institute of Electrical and Electronic Engineers, American Mathematical Society, and others; and the Federal Networking Advisory Committee.

The HPCC Program is driven by the recognition that unprecedented computational power and its creative use are needed to investigate and understand a wide range of scientific and engineering "grand challenge" problems. These are fundamental problems whose solution requires significant increases in computational capability and is critical to national needs. Progress toward solution of these problems is essential to fulfilling many of the missions of the participating agencies. Examples of grand challenges addressed include: prediction of weather, climate, and global change; determination of molecular, atomic, and nuclear structure; understanding turbulence, pollution dispersion, and combustion systems; understanding the structure of biological macromolecules; improving research and education communications; understanding the nature of new materials; and solving problems applicable to national security needs.

The HPCC program nurtures the educational process at all levels by improving academic research and teaching capabilities. Advanced computing and computer communications technologies will accelerate the research process in all disciplines and enable educators to integrate new knowledge and methodologies directly into course curricula. Students at all levels will be drawn into learning and participating in a wide variety of research experiences in all components of this program.

1.1.1   Program Description

The Program consists of four integrated components representing the key areas of high performance computing and communications:

1. *High Performance Computing Systems (HPCS)*. The development of the underlying technology required for scalable high performance computing systems capable of sustaining trillions of operations per second on large problems. Research in very high performance systems is focusing both on

increasing the absolute level of performance attainable and on reducing the cost and size of these very high performance systems in order to make them accessible to a broader range of applications.

2. *Advanced Software Technology and Algorithms (ASTA)*. The development of generic software technology and algorithms and the deployment of the most innovative systems for grand challenge research applications in a networked environment.

3. *National Research and Education Network (NREN)*. The development of a national high speed network to provide distributed computing capability to research and educational institutions and to further advanced research on very high speed networks and applications.

4. *Basic Research and Human Resources (BRHR)*. Support for individual investigator and multidisciplinary long-term research drawn from diverse disciplines, including computer science, computer engineering, and computational science and engineering; initiation of activities to significantly increase the pool of trained personnel; and support for efforts leading to accelerated technology transition.

1.2   DOE ACCELERATED STRATEGIC COMPUTING INITIATIVE

Recently, Los Alamos National Laboratory (LANL) has been focusing on a clustered shared memory multiprocessor architecture approach as part of the ASCI (Accelerated Strategic Computing Initiative).

ASCI planning calls for a new wave of federal support for high-performance computing to stimulate the U.S. supercomputing industry to develop high-performance supercomputers with speeds and memory capacities thousands of times greater than available models and ten to hundreds of times greater than the supercomputers that are anticipated based on trends in development in 1998.

The approach taken by this effort is based on providing the requisite computational resources through clusters of shared-memory multi-processor systems or so-called SMPs. LANL views clustering of SMPS as an economically viable mechanism for achieving the multiple TeraFLOP/s levels of performance required by nuclear stockpile stewardship applications. LANL does not necessarily mean to indicate by SMP the narrower concept of bus-based Symmetric Multi-Processors. Rather, their concept of clusters of SMPs has six components.

1. A hierarchical system of memories and latencies.

2. Multiple high-performance distributed systems.

3. Shared-memory over a significant and economical computational resource with low-latency, high-performance memory access.

4. Multi-processor, supporting a shared-memory environment across multiple processors.

5. Commodity priced SMPs that represent high-end systems of a regular commercial product line (they are not special purpose designs).

6. Commodity priced (third-party) peripherals.

Shared-memory, high-end, microprocessor-based, multi-processor compute servers in the context of this activity are typical examples of the SMP machines.

The emerging clustered SMP computing environment supports a hierarchical distributed computing paradigm covering the full range of the memory and latency hierarchies encountered in clusters. LANL envisions these SMP clusters to be assembled from building blocks whose size is determined by market economics. The approach extends beyond the MPP strategy for large-scale parallelism in which up to thousands of processors are interconnected in a homogeneous flat, distributed memory system and inter-CPU communication is handled by message passing. This ASCI activity will accommodate applications targeted for MPPs but its real advantage lies in the ability to exploit the computational power of multiple system aggregation strategies.

The ideal size for the SMP building block in this approach is determined by a balance of cost effectiveness, reliability, hardware/software scalability, the number of processors per SMP and number of SMPs in the clusters. This balance may change over the lifetime of the effort. The capability on ASCI applications of a single SMP must be as large as economically viable.

Building and delivering a Tera-scale computing resource is a daunting task. Within the context of research and development a well balanced hardware approach will follow the following four notional phases.

1. An initial system for ASCI application code development.

2. Technology refresh as the initial system ages and fruits of the ASCI project become available.

3. One sustained TFLOP/s system with a peak performance at a higher level.

4. Memory upgrade.

The spirit of ASCI is One Program Three Labs. All three National Laboratories will participate in applications development, testing, and applying the resources. This has tremendous impact on the software environment that, of course, is the key to the success of this activity. Effectiveness is expected to occur through system integration and the development of middleware software that allows clusters of SMPs to effectively support a single computation. Hardware transparency is essential in many areas to preserve the investment in the applications beyond the lifetime of the hardware.

The purpose of this activity is to provide a Tera-scale computing environment for the very demanding applications required. A truly usable Tera-scale computing resource requires a software environment that scales as well as the hardware and provides a rich scalable code development environment.

DOE is aware that a Tera-scale software environment is an enormous challenge. Just as in the case of the hardware, the very high performance computing market must significantly stretch to meet these objectives. Therefore, it is LANL's strategy to accelerate software development necessary to achieve these objectives. It is anticipated that activities related to this effort will have at least three general phases.

In the first phase, there will be an intensive applications code development. In addition, it is anticipated that there will be an effort to advance the code development environment and system management tools. In this phase basic ASCI applications development will take place either within an SMP utilizing the SMP programming paradigm (explicit parallelism exploited with threads or implicit parallelism exploited by the compiler) or utilizing message passing via the Message Passing Interface both intra-SMP and intra-cluster. The choice will depend on the best implementation for specific physics packages. ASCI applications will be combinations of many physics packages and will contain both styles of parallelism. On the systems side, it is anticipated that development will be needed to assure robustness and extend the features of the code development environment on a single SMP. This will include compilers, loaders, debuggers, performance analysis tools, gang scheduling, resource management, Distribute Computing Environments, and system administration to name but a few. This is mostly an interactive code development environment.

In the second phase, it is anticipated that the applications will need more batch computing time as well as interactive code development cycles. Development, on the system side, will be extending the cluster notion to provide more of a single system image and operation as a single unit. Basic development of SMP cluster-wide reliability, availability, and serviceability (RAS), gang scheduling, resource management, and performance analysis tools will be extended from an SMP-specific focus to cluster-wide. To give a specific nontrivial example, it is anticipated that a cluster wide process and session identification space will have to be developed to support the single system image, gang scheduling, and resource management.

In the third phase, it is anticipated that most of the exploratory code development environment and the system software development will be completed. It is in this phase that the software productization activities will dominate.

1.3   GRAND CHALLENGES AND SUPPORTING TECHNOLOGY CASE STUDIES

This section incorporates examples of high performance computing and computer communications technologies and several illustrative grand challenge applications in computational science and engineering [1]. The list of grand challenges is too long to allow an example from each possible area, and lack of representation of certain areas does not imply lack of importance. In addition to these examples, there are many important applications of high performance computing in national security areas that are supported by agency mission funding.

The examples that follow were chosen to illustrate the diversity and significance of application areas that have been addressed to date.

- Magnetic recording technology
- High speed civil transports
- Catalysis
- Ocean modeling
- Digital anatomy
- Air pollution
- Venus imaging

1.3.1   Magnetic Recording Technology

The knowledge and information explosion in every aspect of modern life is creating rapidly increasing requirements for efficient storage of information. Currently, the dominant methods for information storage employ magnetic recording media such as the thin magnetic films that are used in computer disk drives and tape systems.

Future advances in magnetic recording technology will underlie new technological advances and scientific breakthroughs, from laptop supercomputers to data collection from earth orbiting satellites downloading terabits of information per day.

One of the goals of magnetic media research is to pack more and more information onto available media. The fundamental problem is to avoid or compensate for magnetic noise in the metallic thin-films used to coat high density disks. The films are crystalline, rather than continuous, and are composed of tiny magnetic grains of cobalt alloys deposited on a substrate of chromium. The grainy, particulate nature of the medium is the fundamental cause of noise.

Researchers are using high-performance computers to study the effects of the two primary magnetic interactions: magnetostatic and exchange. First, all magnetic dipoles (grains) in the films interact with each other via long-range magnetostatic coupling. Second, exchange coupling occurs between neighboring grains when, for example, cobalt diffuses into the intergranular boundaries, causing direct magnetic interaction. Even minuscule exchange coupling, the simulations show, dramatically increases noise.

The smaller that the distinctly separable grains can be made, the higher will be the density at which digital bits can be recorded and retrieved. The key is to devise alloys and fabrication processes to assure nonmagnetic grain boundaries, minimizing exchange-coupling noise. Based on theoretical and computational studies, researchers have recently shown that one billion bits can be packed onto a square inch of disk surface, nearly 30 times the storage density of current media.

1.3.2   High Speed Civil Transports

The United States has initiated a research program to examine the feasibility of a new High Speed Civil Transport (HSCT). This type of aircraft will significantly impact the world's transportation system. Significant technological challenges must be addressed to assure environmental compatibility and to realize the commercial potential. Among the most important of these are: aerodynamic efficiency, advanced materials, and engines with acceptable environmental impact. High performance computing systems are essential to bringing HSCT to reality at the earliest possible date.

Computational fluid dynamics codes will enable the rapid evaluation of alternative HSCT designs for aerodynamic efficiency and stability, as well as predicting the associated sonic boom. In addition, other programs will account for the atmospheric heating of the vehicle and the physical stresses of supersonic flight to provide the necessary information for the development of advanced materials and fabrication of new structures to commercialize HSCTs. High performance computing technologies are crucial to the design and evaluation of advanced propulsion systems that will efficiently and safely power HSCTs while significantly reducing community noise and exhaust emissions harmful to the upper atmosphere.

1.3.3   Catalysis

Catalysts for chemical reactions are currently used in one-third of all manufacturing processes, producing hundreds of billions of dollars worth of products each year. Although most catalysts are still designed empirically, high performance computing, along with state-of-the-art algorithms and physical models, is being used to develop a computer design and analysis capability for commercially important catalytic processes and catalysts. The goal is to reduce the time required to design catalysts and to optimize their properties. Important catalysis applications include enzymes, biomimetic catalysts, microporous solids, and metals.

Many biological processes are catalytically controlled by enzymes, which give a high level of selectivity to a desired reaction over competing reactions. Although enzymes are fundamental in medical science, environmental restoration, and food processing, the way in which enzymes function in their catalytic role is not completely understood for any system.

Enzymatic reactions are attractive because they exhibit an extremely high degree of selectivity to specific reactants and products. However, they also can be limited to chemical and physical environments different from those in many industrial processes. Alternatives to enzymes that may be more useful industrially are biomimetic catalysts, so called because they mimic biological catalysts (enzymes). These catalysts are synthetic materials that generally are much simpler chemically and can function at higher temperatures than their biological counterparts. Computer modeling of these biomimetic catalysts using relatively simple force field models has been crucial in the development of molecules with the correct size and shape properties.

The ocean is the other turbulent fluid in the climate system. By contrast with the atmosphere, the ocean has much smaller spatial structures and much longer time responses. The ocean currents, which transport almost as much heat as atmospheric flows, are only about 50 kilometers across, and ocean eddies that are the counterparts of synoptic weather systems are one-hundredth the size of those in the atmosphere. These oceanic phenomena evolve over periods from years to centuries to profoundly influence the climate. The small space scales and long time scales of the ocean present a massive challenge to high performance computing.

The ocean is particularly important in maintaining the climate of the Northern Hemisphere through an organized system of currents that has been likened to a conveyor belt. Currents near the surface transport heat from the Tropical Pacific all the way to the North Atlantic, where the heat enters the atmosphere through evaporation. The evaporation makes the surface waters heavy enough to sink and return to the Tropical Pacific, where they rise and are freshened by rains and warmed by the sun for another trip to the North Atlantic. However, the entire system is very sensitive to changes in atmospheric conditions, such as might occur from increasing greenhouse gases.

The first global ocean simulation that is capable of barely resolving the eddies and strong currents, including those that act as a conveyor belt, has only recently been accomplished. An integration of just a decade required 3000 processor hours on a gigaflop supercomputer. If multicentury ocean simulations with adequate spatial resolution are to be performed in support of climate prediction, then teraflop machines will be required.

Because of the turbulent nature of oceanic flows, there should be demands for archiving, visualization, and communication that parallel those of the most ambitious atmospheric and hydrodynamic applications. For example, the first global simulation with barely resolved eddies produced more than 250 gigabytes of processed output. A 2-hour animation was made to help understand the resulting phenomena, and the animation required the transfer of gigabytes of model output across high-speed networks.

1.3.4   Digital Anatomy

New computer-based imaging techniques are making it possible to explore the structure and function of the human body with unprecedented accuracy. Clinical imaging methods such as computed tomography, magnetic resonance imaging, and positron emission tomography yield two-dimensional pictures. Reference standard normals, such as the Visible Human project, provide three-dimensional numerical coordinates from which internal as well as external structure can be depicted, rotated, viewed from any angle, and reversibly "dissected."

Automated correlation of images derived from different imaging techniques is a formidable technical problem whose solutions are computation-intensive. These image datasets are also large (e.g., the Visible Human project will generate a four-trillion-byte image library), and high speed networks are the only feasible method for allowing researchers and health professionals to browse such electronic image collections interactively, in a manner analogous to today's interactive browsing of textual databases.

The computer and network tools applied to the challenge of multimodal imaging promise an era of new insight into the anatomy and physiology of human health and disease in the three dimensions of biological structure and the fourth dimension of time.

1.3.5   Air Pollution

Reduction of pollutant and pollutant precursor emissions is costing billions of dollars. The early optimism of the past two decades about the ease of controlling and reducing air pollution, especially ozone, has given way to a realization of the complexity and extreme difficulty of achieving the reductions necessary to attain the health standards in many urban areas. In addition, connections between pollutant media are increasingly being recognized. For example, nitrogen deposition from air pollutants is now thought to be a major contributor to eutrophication of coastal estuaries, adding to the agricultural and urban sources that have been studied and characterized.

Computational models are powerful and necessary tools to study the transport and transformation of pollutants and to provide guidance on the expected effectiveness of emission control strategies. These air quality models must incorporate descriptions of physical, chemical, and meteorological processes that encompass scales from regional (several thousands of kilometers) to local (several kilometers) to adequately represent the production of pollutants. Interactions between many physical processes, such as clouds and precipitation, must be incorporated in the computational descriptions, leading to models that are, of necessity, highly sophisticated and that must be run on supercomputers. Yet, the complexity and completeness of the air quality models must be advanced to more accurately simulate the effectiveness of emissions controls and find effective and efficient solutions to air pollution.

High performance computing will enable additional and more complete descriptions of the physical and chemical processes to be simulated in the air quality models. Advanced visualization techniques will allow the scientists to better study and understand the complex, nonlinear interactions of the atmospheric pollutant system, leading to more informed assessments. High performance computing will enable scientists to explore the complex interactions in the models more quickly and provide the type of guidance needed to reduce air pollution to meet the air quality standards.

1.3.6   Venus Imaging

The Magellan spacecraft has been mapping the surface of Venus since September 15, 1990. Magellan uses a radar instrument to penetrate the thick clouds surrounding the planet to produce 120-meter resolution images of the Venus surface. Venus is the most Earth-like of the planets of the inner solar system, with about the same size, mass, and density. However, the "greenhouse effect" caused by the clouds surrounding Venus results in a surface temperature of about 900°F and pressure at the surface about 90 times that of Earth. Despite these differences, Magellan data for Venus have revealed a wide range of geologic features, from volcanoes to mountain ranges, that are providing insight into how geologic processes on Earth and Venus operate. Magellan has provided images and topographic information for over 90 percent of the Venus surface. The total volume of data returned by Magellan exceeds 3×10^{12} bits or 3 terabits (all previous U.S. planetary missions combined have returned about 9×10^{11} bits).

The Magellan radar data go through several processing steps, starting at a Synthetic Aperture Radar (SAR) processor that turns the echoes returned from the planet's surface into images. The data then go through various image processing steps, where Magellan image strips are combined to form mosaics that can then be color enhanced or combined with topographic data to provide scientists with tools for understanding the geology of Venus. The large data volume associated with Magellan has necessitated new and increasingly more complex methods of data processing, handling and storage that will be significant to future data-intensive projects such as NASA's Mission to Planet Earth. Magellan images provide insights into the physics of volcanism and its significance to planetary evolution.

Using the large parallel supercomputers for producing the rendered images dramatically increases our capabilities. For example, a single image produced on a conventional high performance workstation takes several minutes to complete. Animations, even short ones, typically take weeks to produce. Massively parallel supercomputers hold the promise of being able to store the huge datasets on-line and render animations in real-time.

REFERENCE

1. Grand Challenges 1993: High Performance Computing and Communications, Federal Coordinating Committee for Science, Engineering and Technology, Office of Science and Technology Policy.

Chapter 2

Parallel Supercomputing Concepts

We address three topics in this chapter on parallel supercomputing concepts. The first, supercomputing, defines the term and offers some thoughts about why supercomputing is useful. The second, MIMD versus SIMD, compares and contrasts the multiple-instruction-multiple-datastream and the single-instruction-multiple-datastream architectures, the two major parallel supercomputing paradigms. The third topic, MIMD concepts, examines the implications of having the various processors of a parallel computer executing diverse instruction sequences (unlike the simpler SIMD approach in which all of the processors execute the same instruction sequence in lockstep) simultaneously, and the steps one must take to coordinate these interacting processes.

Generally, if a program can be efficiently implemented on an SIMD architecture, then that is the architecture that should be employed because the programming is simpler and the hardware is less expensive per operation. However, the MIMD method has gradually grown to ascendancy as users and researchers have recognized that MIMD machines address a broader range of algorithms efficiently.

2.1 SUPERCOMPUTING

Supercomputers address the big problems of their time. Because they are expensive, they need a computationally intensive application to warrant their use, but for such problems they are economically justified because they are less costly than their conventional competitors on that class of problems; and, of course, they solve problems that are so big that conventional computers are hopelessly inadequate to address, the so-called grand challenge problems.

Supercomputers achieve their status by virtue of their speed and capacity. In part, their speed is derived from faster circuits; but as Dr. E. W. Martin, Vice

President of Boeing Aerospace, said, as quoted in *Photonics Spectra*, January 1991: "In 40 years, electronic logic-element switching speeds have increased only three orders of magnitude; however, computer speeds have increased by nine orders of magnitude. The key to this gain is architectures that provide increased parallelisms."

2.2   MIMD VERSUS SIMD

Parallel processors fall mainly into two general classes as described in Table 2-1. The fundamental distinction between the two classes is that one class is MIMD (multiple instruction multiple data), whereas the other is SIMD (single instruction multiple data). This distinction can be summarized as follows.

TABLE 2-1

TWO GENERAL CLASSES OF PARALLEL PROCESSING	
MIMD	**SIMD**
Relatively few powerful processors	Many simple processors
Control level parallelism that assigns a processor to a unit of code	Data level parallelism that assigns a processor to a unit of data
Typically either distributed memory or shared memory; can have memory contention	Typically distributed memory; can have data communication problem
Needs good task scheduling for efficiency	Needs good processor utilization for efficiency
Multiple Instruction Multiple Data	
Each processor runs its own instruction sequence Each processor works on a different part of the problem Each processor communicates data to other parts Processors may have to wait for other processors or for access to data	
Single Instruction Multiple Data	
All processors are given the same instruction Each processor operates on different data Processors may "sit out" a sequence of instructions	

In MIMD architectures several processors operate in parallel in an asynchronous manner and generally share access to a common memory. Two features are of interest to differentiate among designs: the coupling of processor units and memories, and the homogeneity of the processing units.

In tightly coupled MIMD multiprocessors, the number of processing units is fixed, and they operate under the supervision of a strict control scheme. Generally, the controller is a separate hardware unit. Most of the hardware controlled tightly coupled multiprocessors are heterogeneous in the sense that they consist of specialized functional units (e.g., adders, multipliers) supervised by an instruction unit that decodes instructions, fetches operands, and dispatches orders to the functional units.

In SIMD architectures a single control unit (CU) fetches and decodes instructions. Then the instruction is executed either in the CU itself (e.g., a jump instruction) or it is broadcast to a collection of processing elements (PEs). These PEs operate synchronously, but their local memories have different contents. Depending on the complexity of the CU, the processing power and the addressing method of the PEs, and the interconnection facilities between the PEs, we can differentiate among various subclasses of SIMD machines, for example, array processors, associative processors, and processing ensembles.

2.2.1 Fine Grain Versus Coarse Grain

There is some confusion about the terms fine grain and coarse grain as applied to parallel computers. Some use the terms to characterize the power of the individual processors. In this sense a fine-grain processor is rather elemental, perhaps operating on a one-bit word and having very few registers. This is contrasted with powerful processing elements in coarse-grain computers, each a full computer in its own right with a 32- or 64-bit word size.

Others have used the terms to differentiate between computers with a small number of processors and those with a large number. More recently those with a small number have been termed simply "parallel," whereas those with a large number are termed "massively parallel." The dividing line between these classes has been set at 1000 so that computers with more than 1000 processors are termed massively parallel.

2.2.2 Shared Versus Distributed Memory

Another distinction between MIMD and SIMD architecture classes is whether the memory is shared or distributed. In shared memory designs all processors have direct access to all of the memory; in distributed memory computers each processor has direct access only to its own local memory. Typically MIMD computers use shared memory, whereas SIMD computers use distributed memory.

2.2.3 Interconnect Topology

The final major distinguishing characteristic among current parallel computers is the topology of the interconnections, that is, which processors have direct interconnection with which other processors. Diagrams of various examples are

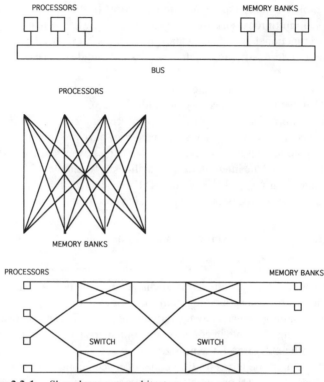

Figure 2.2-1.   Shared memory architecture.

shown in Figures 2.2-1 and 2.2-2. Figure 2.2-1 shows variations in connectivity for the shared memory class of architecture. Figure 2.2-2 illustrates five kinds of connectivity for the distributed memory architecture class.

Interconnect topology affects the rate at which internode communication occurs. Generally this is measured as "bisectional bandwidth." For discussions of bisectional bandwidth see Sections 5.1 and 5.3. It is also important, in assessing internode communication, to understand message flow control methods that are commonly divided into circuit switched methods and packet switched methods. These are discussed in Section 5.3. The use of prefetching remote data to avoid delays associated with long message paths is discussed in Section 5.5.

2.3   MIMD CONCEPTS

It is possible to create a vast variety of MIMD parallel programs with just five library functions: one for creating processes; another for destroying processes; a

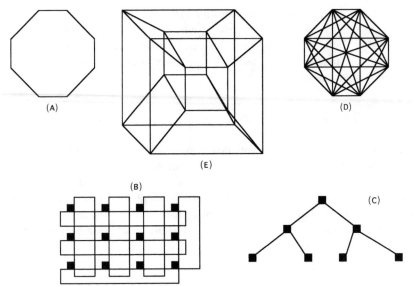

Figure 2.2-2. Distributed memory architecture. (A) Ring; (B) mesh; (C) tree; (D) fully interconnected; (E) hypercube.

third for sharing memory; and two more for interprocess synchronization. The complexity of parallel programming in a real application is owing to the interactions among these primitives [1].

The five functions are the following.

1. Forks, to create a parallel process;

2. Joins, to destroy a parallel process;

3. Memory shares;

4. Memory locks; and

5. Barriers, which are used to synchronize parallel programs.

One way to spread a single calculation among several different processors is to make each part of the calculation into an independent program, or appear to be an independent program to the operating system. In reality, of course, they are not independent because they share intermediate results and they may share data. In MIMD parallel programming, these "independent" programs are termed processes, a generalization of the program concept and the logical foundation of shared memory multiprocessing.

In a time-shared uniprocessor, two users can be running programs at the same time. For parallel processing, this scenario is extended in three ways: The two users are one person, the programs are processes that can communicate, and the two programs do not share a single processor, but instead each has its own dedicated processor. Naturally this extension from time-shared uniprocessing applies to three, four, and more users. A parallel program may have hundreds or thousands of processes.

2.3.1   Shared Memory

When a calculation has been subdivided among two or more processes, each must be able to carry out its portion of that calculation. For this to be the case, each process must have some part of the total memory that is private to itself to be certain that the value it previously stored will still be there when it goes back to get it. However, there must also be part of memory that all processes can access. These memory locations that can be accessed by all of the processes are termed shared.

2.3.2   Forks

Processes are created by the operating system. The user can command the operating system to create a new process by calling the library function process_fork. Some libraries call this function by a different name; if the function is not in a library at all, the function can be constructed based, for example, on UNIX® system calls. (UNIX is a registered trademark of AT&T.)

The function process_fork is used as follows.

```
integer id, nproc, process_fork
id = process_fork (nproc)
```

The function process_fork (nproc), when called from an executing process, creates nproc-1 additional processes. Each additional process is an exact copy of the caller. The original process, which is called process_fork, is still running, so there are now nproc processes in all. Each of the nproc processes has a private copy of each of the variables of the parent, unless the variable is explicitly shared. The process that made the call is termed the parent, and the newly created processes are termed the children. If nproc = 1 in the original call, no new processes are created, that is, there are no children.

Although the process_fork function is invoked by only one process, it returns nproc times to nproc identical processes of which nproc-1 are new. After the return, each process continues executing at the next executable instruction following the process_fork function call.

The function process_fork returns an integer, called the process-id, which is unique to each process. It returns 0 to the parent and the integers 1,2,3 . . . nproc-1 to the children, each child getting a different number. That each child gets a different number allows the different processes to identify themselves and do a different calculation.

Joins. The join is the opposite of a fork. It destroys all the children, leaving only the parent running. The join is executed by the following statement.

```
call process_join()
```

The subroutine process_join has no arguments, so the parentheses are optional. The user's library may have a program with the functionality of process_join but, with a different name.

If a child makes the process_join call, it is destroyed. If the calling process is the parent (id = 0), it waits until all the children are destroyed, then it continues with the statement following the call to process_join. When a process is destroyed, the memory it occupied is released for other uses. The operating system no longer has a record of the destroyed process. There is no way to access the data of the destroyed process.

Randomly Scheduled Processes. In analyzing shared memory parallel programs it is essential to assume that all processes are randomly scheduled. Any process can be idled and restarted at any time. It can be idled for any length of time. It is incorrect to assume that two processes will proceed at the same rate or that one process will go faster because it has less work to do than another process. The operating system can intervene at any time to idle a process.

This random scheduling of processes can cause the order of the output of different processors to be different in each run. It is also possible for processors to incorrectly alter something in shared memory so that the final value of a shared variable will be different depending on the order in which the variable is altered.

Self-Scheduling and Spin-Locks. One common way to share a calculation among nproc processors is loop splitting. The general form of loop splitting follows.

```
      do 1 i = 1 + id, n, nproc
            (work)
 1 continue
```

On a sequential uniprocessor the loop would start with i = 1 and end n passes later with i = n. On a multiprocessor with nproc processors, process 0 would perform the work of loops i = 1, i = 1 + nproc, i = 1 + 2 * nproc, . . . , whereas process 1 would perform the work of loops i = 2, i = 2 + nproc, i = 2 + 2 * nproc, . . . , etc.

The work section of this loop can be long for some values of i, but short for other values of i. Consider an example where several processes are exploring a maze. Some processes may reach a dead end in a short time; others may take a long time. Some processes may explore short paths, whereas others may explore long paths.

The alternative is self-scheduling; each process chooses its next index value only when it is ready to begin executing it. Self-scheduling means that a processor can take onto itself the next job in the queue whenever it is ready to do so. Self-scheduling allows some processes to execute only a few iterations, whereas others execute many. This permits the ensemble of processes to finish more closely at the same time, keeping all processors in use. Self-scheduling is invoked by the following.

```
call shared (next_index, . . .)
```

In the call to shared, next_index has the next available value of the index i that will be used by the process. After the process assigns the value of the next_index to i, decrements next_index so that the next process can assign the next lower value of the index. The intention of the program is to allocate the values of i to processes on a first-come, first-served basis.

Difficulty arises in the event of contention, that is, if two or more processes try to alter the value of next_index at the same time. If process 0 assigns i = next_index and process 1 assigns i = next_index before process 0 can decrement next_index, a program can perform incorrectly. To prevent this, the assignment and the decrement operations must be treated as a unit so that there is no interruption and, once started, both operations complete. This section of operations is therefore "protected" in that while a process is executing protected code, all other processes are locked out.

In order to enforce protection, it is necessary that one process communicate with all other processes, that it is in a protected portion of the program, and that all other processes must not interfere. Such communication is an example of synchronization.

In the case of self-scheduling, the required software structure for implementing this synchronization is the spin-lock. The spin-lock is normally supplied by the operating system.

The use of the spin-lock requires three function calls and a shared variable. The statement

```
spin_lock_init (lock_word, lock_condition)
```

initializes the spin-lock. A spin-lock has only two conditions, locked and unlocked. The integer lock_word is shared by all the processes and describes the state of the spin-lock. For example, when the state of the spin-lock is unlocked, then lock_word is zero, which means that there is no process executing instructions in a protected region of the code. When lock_word is one, some process is in a protected region and the spin-lock is locked. The lock_condition value in the call to initialize indicates whether the initial condition is for the spin-lock to be locked or unlocked.

Besides the initialization, there are two functions that actually perform the work of protecting a group of instructions.

```
call spin_lock (lock_word)
call spin_unlock (lock_word)
```

A process calls spin-lock when it is about to enter a protected region of its code. When called, the function first checks to see whether the lock is unlocked. If it is, the lock is locked and the process proceeds into the protected region. However, if the function finds the lock is already locked, then the calling process must wait until the lock becomes unlocked. Thus, the process that seeks to enter a protected region cannot proceed until the other process that caused the lock to go to the locked condition completes its protected code and unlocks the lock. It is this behavior that gives rise to the name spin-lock. The call to spin-lock puts the process into a loop

```
100 continue
        if (lock_word.eq.1) go to 100
```

that spins around and around until lock_word has the value 0, indicating that the lock has been unlocked. When this happens, only one of the waiting processes can enter the protected region; the remaining processes keep spinning while they wait their turn. The order in which waiting processes enter the protected region is undefined. When properly employed, a spin-lock eliminates contention, that is, only one process is able to update a shared variable at a time.

The spin-lock is the most basic synchronization mechanism. Practical parallel programming requires a generous use of locks. However, an inefficient implementation of locks can make a parallel program run slower than the sequential version.

Barriers.   A barrier causes processes to wait and allows them to proceed only after a predetermined number of processes are waiting at the barrier. It is used to ensure that one stage of a calculation has been completed before the processes proceed to a next stage, which requires the results of the previous stage. Barriers are used to eliminate race conditions, in which the result of a calculation depends on the relative speed at which the processes execute.

REFERENCE

1. Brawer, S., Ed., Portions of the text taken from *Introduction to Parallel Programming*, Academic Press, San Diego, CA, 1989.

Chapter 3

History

This chapter examines the evolution of parallel supercomputers over the last thirty years. The intent is to acquaint the reader with the historical forerunners of today's state-of-the-art machines, and the ideas that evolved along with the technology.

This material contributes to a deeper understanding of the field; only through a knowledge of the evolution of parallel supercomputers may one appreciate today's capabilities and become an informed user of this technology.

Seven computers that were the supercomputers of their day are presented in some detail. The Illiac IV and the Connection Machine-2 are given emphasis as pivotal events in the progress of parallel supercomputing.

3.1 PEPE (PARALLEL ELEMENT PROCESSING ENSEMBLE)

PEPE was a special-purpose "attachment" to a general-purpose computer called the host. It was housed in the Ballistic Missile Defense Agency quarters in Huntsville (Alabama) in the 1970s. PEPE's main application was radar processing. A block diagram of the system is shown in Figures 3.1-1, 3.1-2, and 3.1-3. The host is a CDC 7600 and the test and maintenance station is controlled by a Burroughs B1700. PEPE itself consists of a control system and up to 288 processing elements (PEs). PEs can be added or disconnected without impeding the normal operation of the system. One reason for this flexibility is that, unlike most other SIMD systems, there is no direct connection between PEs. If data has to be transferred between PEs it will have to be routed through the host.

A PE consists of three units: the arithmetic unit (AU), correlation unit (CU), and associative output unit (AOU) sharing an element memory (EM). The three types of units are under the global control of three control units: arithmetic control unit (ACU), correlation control unit (CCU), and associative output control

TABLE 3-1

PEPE CHARACTERISTICS	
Class	
Single instruction stream, multiple data stream Parallel processing Associative data match for input Associative search for output Simultaneous input, output, compute Conventional floating point, integer, logical instructions plus associative match and search instructions	
Data processing speed	
Arithmetic control unit	1 MIP average times number of active elements
Correlation control unit	5 MIP average times number of active elements
Associative output control unit	5 MIP average times number of active elements

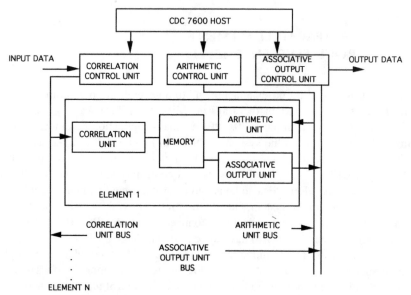

Figure 3.1-1.   PEPE architecture.

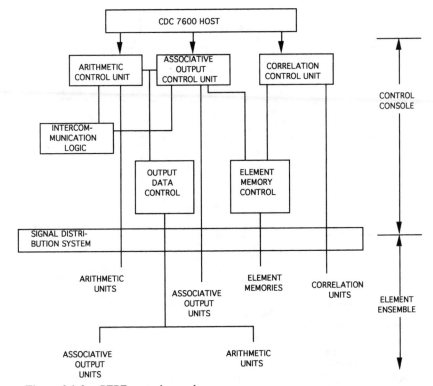

Figure 3.1-2.   PEPE control console.

unit (AOCU), which are part of the control system. Not shown in Figure 3.1-3 are units in the control system for output data control and element memory control to resolve conflicts in data access, the interconnection logic between ACU, CCU, and AOCU, and an I/O unit to connect to the host.

The PEs perform most of the computational work. For example, in a radar processing application each PE would have the responsibility of an object in the sky, maintaining a data base for this object in its EM. Arithmetic computations such as track updating and prediction can be performed, in parallel, in each AU. Each PE can be enabled/disabled by setting a control flip-flop. The AUs execute under control of the ACU, which sends microprogrammed sequences to them. Individual AUs do not have a stored program of their own.

Inputs to the PEs are controlled by the CUs under the supervision of the CCU. For example, information on a new object can be broadcast to all CUs at the same time. Each CU will correlate the new coordinates with predictions performed by the AU in the same element. The new information can then be input to the CU whose data correlates, or to the first empty element if there were no correlations. It is this type of processing that gives to PEPE its associative level since

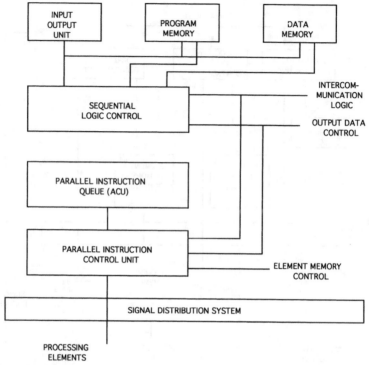

Figure 3.1-3.   PEPE arithmetic control unit.

the broadcast data can be viewed as an argument to be matched by the CUs acting as memory cells.

The AOUS, under control of the AOCU in an enable/disable fashion similar to the ACUs, provide data for the radar connected to the host. The data has to be ordered on an object by object basis and this ordering is performed through an associative maximum-minimum search. The three units in a PE can operate concurrently. Since PEPE can have 288 elements, up to 864 operations can be in execution at the same time. In order to alleviate the loads on the input and output buses connecting the global control units and the PEs, most programs for the latter are loaded at initialization time with very few parameter modifications during execution.

The three global control units have the responsibility of the control flow. Each control unit has its own program and data memories. They can communicate with each other. A program is a mix of instructions executed either in the global control unit or broadcast to all PEs.

This architecture is well suited for parallel tasks with low intertask communications requirements. The associative processing and the fact that the system can be in operation with a variable number of PEs are interesting and original features.

3.2   STARAN

During the 1970s, Goodyear Aerospace Corporation produced an associative processor called STARAN. The last model, Series E, had the same general architecture as its predecessors. (For a discussion of ASPRO, a newer version of STARAN, see Section 4.4.) The main improvement was an expanded memory capacity. The salient features of STARAN were that it could simultaneously perform search, arithmetic, and logical operations on either all or selected words of its memory. The memory would be accessed on a word-by-word basis or in a bit-slice manner. A processing element was associated with each memory word of 256 bits. The words were grouped into arrays of 256 words. STARAN could have from 1 to 32 array elements.

Because STARAN's designers felt that each system should be custom built for its users, the overall configuration has the form of Figure 3.2-1, with STARAN interconnected through a custom interface unit to peripherals and other computers. The parallel I/O channel, PIO, can transfer in parallel up to $32 \times 256 = 8192$ words, that is, it has a direct connection to all memory words. The STARAN system is shown in more detail in Figure 3.2-2.

An array element is shown in Figure 3.2-3. It consists of a 256×256 multi-dimensional access memory (in STARAN E, this was expanded to several planes of 256×256 bits), 256 processing elements (one per memory word), and a flip (also called scramble/unscramble) interconnection network between the memory and the processing elements. The flip network was designed in such a way that words and bit-slices, as well as other templates, could be implemented using conventional RAM chips.

STARAN is well suited for applications that require parallel processing at the bit level (e.g., data base management, air traffic control) and word (or bit-group) level (e.g., arithmetic operations). Furthermore, the inclusion of the permutation network facilitates the efficient programming of functions such as the

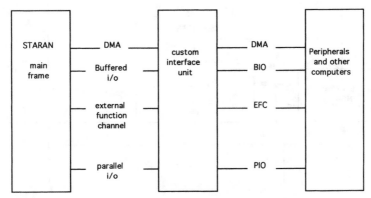

Figure 3.2-1.   STARAN system configuration.

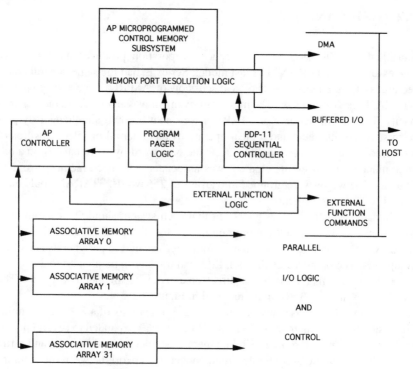

Figure 3.2-2. STARAN system.

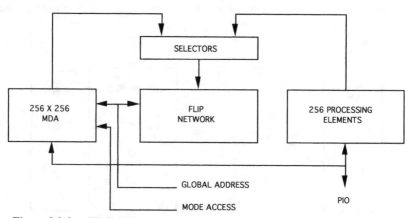

Figure 3.2-3. STARAN's array element.

fast Fourier transform. STARAN cannot be used as a stand-alone facility but must be part of a complex where it only performs parallel tasks.

3.3   SOLOMON I

A third early machine, SOLOMON I, built at Westinghouse by a team led by Daniel Slotnick, is generally considered the forerunner of the Illiac IV. A diagram of a processing element is shown in Figure 3.3-1. A partial block diagram of the control unit of the SOLOMON I is shown in Figure 3.3-2.

3.4   ILLIAC IV—THE FIRST SIMD SUPERCOMPUTER

The Illiac IV was the first large-scale array computer [1]. As the forerunner of today's advanced computers, it brought whole classes of scientific computations into the realm of practicality. Conceived initially as a grand experiment in computer science, the revolutionary architecture incorporated both a high level of

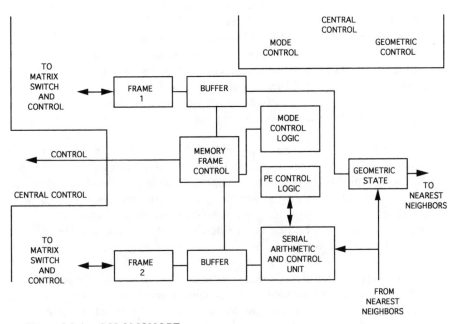

Figure 3.3-1.   SOLOMON I PE.

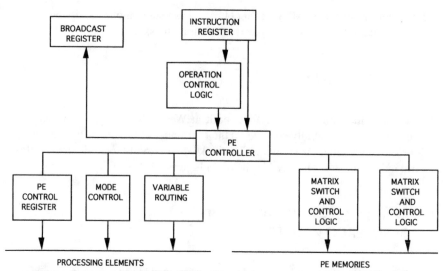

Figure 3.3-2.   SOLOMON I sequencer (partial block diagram of SOLOMON I control unit).

parallelism and pipelining. After a difficult gestation, the Illiac IV became operational in 1975.

The Illiac IV consisted of a single control unit that broadcast instructions to 64 processing elements operating in lock step. Each of these processing elements had a working memory of 2 K 64-bit words. The main memory of the Illiac was implemented in disk with a capacity of 8 million words and with a transfer rate of 500 megabits per second. Arithmetic could be performed in 64-, 32-, or 8-bit mode; on algorithms well suited to the parallel architecture, the Illiac performed at a rate of 300 million instructions per second. Although it used electronics from the late 1960s, for certain classes of important problems, the Illiac was the fastest computer of its time.

The Illiac IV story begins in the mid-1960s. Then, as now, the computational community had requirements for machines much faster and with more capacity than were available. Large classes of important calculational problems were outside the realm of practicality because the most powerful machines of the day were too slow by orders of magnitude to execute the programs in plausible time. These applications included ballistic missile defense analyses, reactor design calculations, climate modeling, large linear programming, hydrodynamic simulations, seismic data processing, and a host of others.

This demand for higher speed computation began in this time frame to encounter the ultimate limitation on the computing speed theoretically achievable with sequential machines. This limitation is the speed at which a signal can be propagated through an electrical conductor.

Designers realized that new kinds of logical organization were needed to break through the speed of light barrier to sequential computers. The response to this need was the parallel architecture. It was not the only response. Another architectural approach that met with some success is overlapping or pipelining, wherein an assembly line process is set up for performing sequential operations at different stations within the computer in much the way an automobile is fabricated. The Illiac IV incorporates both of these architectural features.

3.4.1   The Design Concept

The Illiac IV computer was the fourth of a series of advanced computers designed and developed at the University of Illinois (which accounts for the origin of its name). Its predecessors include a vacuum tube machine completed in 1952 (11,000 operations per second), a transistor machine completed in 1963 (500,000 operations per second), and a 1966 machine designed for automatic scanning of large quantities of visual data. The Illiac IV was a parallel processor in which 64 separate computers worked in tandem on the same problem. This parallel approach to computation allowed the Illiac IV to achieve up to 300 million operations per second.

The father of the Illiac IV was Professor Daniel Slotnick who conceived that machine in the mid-1960s. The development was sponsored by the Defense Advanced Research Projects Agency. Subsystems for the Illiac were manufactured in a number of facilities throughout the United States. These subsystems were then shipped to the Burroughs Corporation in Paoli, PA for final assembly. The Illiac was delivered to the NASA Ames Research Center south of San Francisco in 1971.

The logical design of the Illiac IV was patterned after the Solomon computers. Prototypes of these were built in the early 1960s by the Westinghouse Electric Company.

For comparison, the logical structure of a conventional sequential computer is illustrated in Figure 3.4-1, whereas Figure 3.4-2 shows the architecture of the SIMD machine.

In the particular case of the Illiac IV, each of the processing element memories had a capacity of 2048 words of 64-bit length. In aggregate, the processing element memories provided a megabyte of storage. The time required to fetch a number from this memory was 188 ns; however, because additional logical circuitry was needed to resolve contention when two sections of the Illiac IV accessed memory simultaneously, the minimum time between successive operations was somewhat longer.

In the execution of a program it was often necessary to move data or intermediate results from one processor to another. Routing paths for this purpose were provided as shown in Figure 3.4-3. One way of regarding this interconnection pattern is to consider the processing element as a linear string numbered from

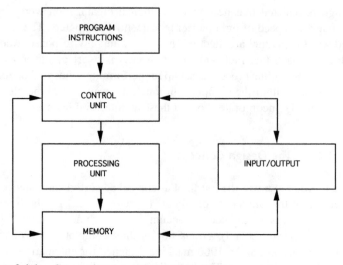

Figure 3.4-1.   Conventional computer architecture.

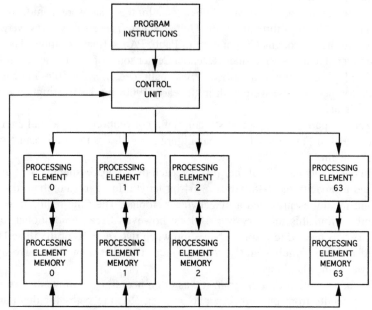

Figure 3.4-2.   Parallel organization of a SIMD computer.

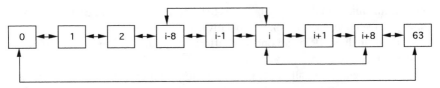

Figure 3.4-3.   Illiac IV routing paths.

0 to 63. Each processor was provided a direct data path to four other processors, its immediate right and left neighbors and the neighbors spaced eight elements away. So, for example, processor 10 was directly connected to processors 9, 11, 2, and 18. This interconnection structure was wrapped around, so processor 63 was directly connected to processor 0. To transfer values among processors not directly connected, multiple routing steps were required. For example, to move a number from processor 9 to processor 18 it was first moved to processor 17 and then to processor 18.

3.4.2   Implementation Difficulties

Illiac was funded by the U.S. Department of Defense Advanced Research Projects Agency (ARPA) through the U.S. Air Force Rome Air Defense Center. However, the entire project was not only conceived, but to a large extent managed, by academicians at the University of Illinois. Finally, the system hardware was actually designed and built by manufacturing firms—Burroughs acted as the overall system contractor; key subcontractors included Texas Instruments and Fairchild Semiconductor.

When headlines in the *Daily Illini* [2], proclaimed, "Department of Defense to employ UI computer for nuclear weaponry," tensions rapidly escalated—not only between University of Illinois students and the faculty and school administration, but also between the parties directly involved in the Illiac project itself.

The end results this pioneering urge had on computer hardware were impressive; Illiac IV was one of the first computers to use all semiconductor main memories; the project also helped to make faster and more highly integrated bipolar logic circuits available (a boon to the semiconductor and computer industries, this development actually proved a disaster for Illiac IV); in a negative but decisive sense, Illiac IV gave a death blow to thin-film memories; the physical design, using large, 15-layer printed circuit boards, challenged the capabilities of automated design techniques.

When Illiac IV was delivered to its final home at the NASA Ames Research Center in California in the spring of 1972, the question in the mind of Dr. Pirtle, former Director of NASA's Institute for Advanced Computation, was whether or not the machine could actually be made to perform useful work. By the following summer, the educated outlook was positive.

Then, in June 1975, a concerted effort began to check out Illiac fully and make it operational.

The system eventually operated from Monday morning to Friday afternoon, including 60 to 80 h of good, verified uptime for the users, along with 44 h of maintenance and downtime.

From the system software standpoint, Illiac IV was quite rudimentary. There was almost no operating system. A user took hold of the machine, ran his or her problem, and then let it go; the next user did the same. No shared use of Illiac's 64 processing elements was provided. In smaller computers that surrounded Illiac's control unit and processing elements, there was more complex software that formed a queue of users waiting to get at the big machine and allowed them to perform nonarithmetic "companion" processes. However, the actual Illiac operating software itself was very simple, capable of such basic operations as monitoring input/output and loading data into the processing element memories. An operating system, along with two Illiac IV languages called TRANQUIL and GLYPNIR, was written at the University of Illinois beginning in 1966. This effort amounted to perhaps a dozen man-years of programming. Later, when the system was moved to California and connected to the ARPA network, it was decided that entirely new system software was needed, since PDP-10 and PDP-11 computers were used—in place of the original B6500 machine—to connect Illiac IV to the outside world.

There, the NASA Ames users decided to write a new Illiac IV language, which would be called CFD, to efficiently communicate problems involving the solution of partial differential equations to the big machine. This was accomplished with approximately 2 man-years of programming effort.

These equations were important to the NASA Ames users, who took up about 20% of Illiac IV operating time solving aerodynamic flow equations.

The remaining 80% of Illiac IV time was taken up by a diverse, and often anonymous, group of users, many of whom used the GLYPNIR language.

A giant computer should be useful for tackling giant computing problems, and that was pretty much the story of Illiac IV applications programs. Beyond the NASA Ames aerodynamic flow problems, users of the big computer ran several small weather-prediction and climate models with improved and larger models still under development.

Several types of signal processing computations, including fast Fourier transforms, were a regular part of Illiac IV's diet, and a large-scale experiment with real-time data was performed. Other applications problems that actually found their way to Illiac IV include beam forming and convolution, seismic research, radiation transport for fission reactors, and linear programming software.

3.4.3   I4 System

Illiac was a parallel processor. It consisted of a control unit (CU), 64 processing elements (PE), 131,072 words of core memory, and 15,974,400 words of disk memory. The control unit had access to all of core memory. Its basic cycle time was 60 ns. However, greater processing power was achieved through the simultaneous execution of an instruction in each of the 64 processing elements.

The control unit fetched and decoded all instructions. After decoding, some instructions were broadcast for execution in the processing elements while others were executed in the control unit. The arithmetic capability of the control unit was limited to 24-bit two's complement addition and subtraction, masking, and comparison for use in branching. The control unit had no floating point capability. One operand at a time was processed by the control unit. The control unit also initiated data transfers between core and Illiac disk.

The processing power of Illiac resided in 64 identical processing elements. Each PE executed instructions broadcast from the CU. Although each PE had its own index registers and memory to operate on, all 64 PEs always executed identical instructions in lock-step. Each PE had direct access to 2048 words of core memory.

There were three data paths available for communication among PEs and between the PEs and the control unit (CU). First, the CU could access all of core, so it could load a word from one processing element memory (PEM) and either use it or store in another PEM. This method of communication was both simple and flexible, allowing for any data movement desired, but, since only one word at a time was transferred, it was relatively slow compared to the two other methods available.

Second, the CU could communicate with all PEs by broadcasting the same word to all PEs simultaneously. This method was faster than the first, since 64 words were transmitted at once, but provided only a limited form of communication.

Third, the PEs could communicate with each other via the ROUTE instruction, which transferred the contents of a register in each PE to another PE.

The primary memory used by Illiac was a disk memory with capacity approximately 100 times that of core memory. One page (1024 words) of memory was the minimum amount of data that could be transferred between core and disk. Although the bandwidth between core and disk was $5. \times 10^8$ bits/s, the average access time to a particular spot on the disk was 20 ms. This relatively long access time (compared to an 80-ns clock time in 64 parallel processors) necessitated careful planning of disk usage. The number of disk transfers was kept to a minimum to avoid waiting for disk accesses.

Table 3-2 contains timings for the execution of a commonly used vector operation on the CDC 7600, CDC Star 100, Illiac IV and CRAY 1.

Table 3-2

		5	10	50	100	500	1000
Vector Operation Timings $V(I) = (A(I) + S1)*S2$							
FORTRAN	CDC 7600	2.1	2.4	2.7	2.8	2.8	2.8
RDALIB	CDC 7600	—	—	10.0	10.6	11.1	11.1
STAR 100	64 bit	1.1	2.0	8.2	13.2	25.5	28.9
STAR 100	32 bit	1.3	2.5	11.3	20.3	56.1	71.8
ILLIAC IV	64 bit	2.40	4.80	24.0	26.55	39.7	41.0
ILLIAC IV	32 bit	1.47	2.937	14.7	29.37	60.1	67.21
CRAY 1		—	—	59.3	60.2	63.9	64.4

The units used to measure performance are MFLOPS, or millions of floating point operations per second.

References

1. Hord, R. Michael, *Illiac IV—The First Supercomputer,* Computer Science Press, Rockville, MD, 1982.
2. *Daily Illini,* Jan 6, 1970.

3.5   BUTTERFLY

The Butterfly Parallel Processor (BB&N, Cambridge, MA) was composed of processors with memory and a novel high performance Butterfly Switch interconnecting the processors [1]. One processor and its memory were located on a single board called a Processor Node. All Butterfly Processor Nodes were identical. I/O connections could be made to each Processor Node, making I/O configuration very flexible. Collectively, the memory of the Processor Nodes formed the shared memory of the machine. All memory was local to some Processor Node; however, each processor could access any of the memory in the machine, using the Butterfly Switch to make remote references. From the point of view of a program running on one node, the only difference between references to memory on its local Processor Node and memory on other Processor Nodes was that remote references took a little longer to complete. Typical memory referencing instructions accessing local memory took about 2 microseconds (μsec) to complete, whereas those accessing remote memory took about 6 μsec. The speeds of the processors, memories, and switch were balanced to permit the system to work efficiently in a wide range of configurations.

The shared-memory architecture of the Butterfly parallel processor, together with the firmware and software of the Butterfly's operating system, Chrysalis™, provided a program execution environment in which tasks could be distributed among processors with little regard to the physical location of data associated with the tasks. This greatly simplified programming the machine, and permitted effective utilization of the multiple processors over an impressive variety of applications.

The Butterfly architecture scaled in a flexible and cost-effective fashion to meet a range of processing and memory requirements. A Butterfly system could be configured with from 1 to 256 Processor Nodes to achieve computing power matched to the needs of a particular application. Memory, switch, and I/O capacity scaled with the number of processors to maintain overall system balance and to prevent potential performance bottlenecks. A large Butterfly configuration achieved its processing power by using many low-cost processors working cooperatively. This was in contrast to large mainframes, which used costly technology to achieve their high processing capacity. Each Processor Node was capable of executing 500,000 instructions per second and was usually configured with 1 MByte of memory (expansion memory was available to increase this to 4 MByte). The bandwidth through each processor-to-processor path in the Butterfly Switch was 32 megabits/sec. Thus, a 256 processor machine had a raw processing power of 128 MIPS, 256 MBytes of main memory (1024 MBytes when fully configured with expansion memory), and an interprocessor communication capacity of 8 gigabits/sec.

Another feature of the Butterfly architecture was its relative insensitivity to component failures. The machine could function without one or more of its Processor Nodes. For example, a 128 processor machine would function with 98% of its capacity even if three of its processors had failed or been removed.

In summary, the Butterfly Parallel Processor was modular, powerful, reliable, cost-effective, and easy to program. These characteristics of the Butterfly system made it attractive for a wide range of applications. It was well suited for computationally intensive applications, such as modeling and simulation. It could economically undertake the solution of compute-bound problems in the physical sciences, such as those found in molecular mechanics and crystallography, and its capabilities were well matched to the needs of the finite element method widely used in structural mechanics. Relatively small configurations (1–16 processors) were used in data communications applications as Internet gateways, as satellite channel controllers for packet communication systems, and as multiplexing concentrators and demultiplexers for voice terminals. Larger configurations were used for research in image processing and computer vision. The Butterfly computer was used for circuit simulation to support VLSI design. Symbolic processing and AI applications using the machine were developed.

The Butterfly architecture overcame the problems that limited the expandability of bus and crossbar interconnection architectures. Because a bus has a

fixed capacity, bus architectures must, in effect, be designed for a given performance level. As a result, bus architectures are relatively expensive in small configurations. (Since each bus interface must operate at the full bus speed, they must
be high speed to offer high performance for larger configurations). Also, they
cannot be effectively used when expanded to configurations that exceed the fixed
capacity of the bus. Crossbar architectures, on the other hand, have a matrix of interconnection points, called cross points, for connecting system elements. Unlike
a bus architecture, the capacity of a crossbar architecture can grow with the number of system elements simply by including enough cross points. Unfortunately,
the number of cross points required is proportional to the square of the number of
processors. As the size of a configuration grows, the cost of the crossbar comes to
dominate the cost of the system, destroying its cost-effectiveness.

The Butterfly was a homogeneous multiprocessor. Although some nodes
may have attached I/O devices, Processor Nodes were basically identical. As a result, every processor was capable of performing any application task. The uniformity of the Butterfly architecture simplified programming, since programmers
did not need to concern themselves with allocating certain tasks to specific
processors. Furthermore, this uniformity made it relatively straightforward to
write application software that was insensitive to the number of available processors.

The Butterfly system had three major components: Processor Nodes, the
Butterfly Switch, and I/O hardware. Each is described in the following. The
switch served to interconnect the Processor Nodes. Figure 3.5-1 illustrates
the Processor Nodes and switch for a 16-processor Butterfly configuration. The
figure shows the switch as a cylinder, with each Processor Node connected to
the switch cylinder through an entry port. Data sent by a Processor Node enters
and is routed through the switch, and exits at the destination Processor Node.

The Butterfly Processor Node contained a Motorola MC68000 microprocessor, at least 1MByte of main memory, a microcoded co-processor called the
Processor Node Controller (PNC), memory management hardware, an I/O bus,
and an interface to the Butterfly Switch. Figure 3.5-2 is a block diagram of the
Processor Node.

All user application software ran on the Motorola MC68000. Although it was
not an absolute requirement of the architecture, all code executed by the
MC68000 resided in local memory.

The Butterfly Switch used techniques similar to packet switching to implement high-performance, reliable, and economical interprocessor communication.
The switch was a collection of switching nodes configured as a "serial decision"
network. Its topology was similar to that of the Fast Fourier Transform Butterfly,
hence the name Butterfly Switch.

Figure 3.5-3 illustrates the 16 processor Butterfly system from Figure 3.5-1
in more detail. The cylinder has been cut down the middle of the Processor
Nodes, flattened, and drawn with half of each Processor Node on each side. Each

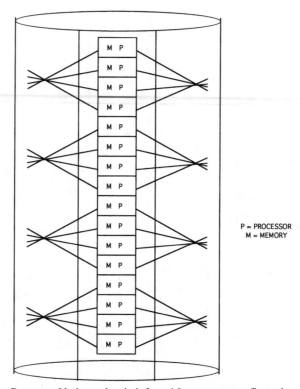

Figure 3.5-1.   Processor Nodes and switch for a 16-processor configuration.

Butterfly switching node was a 4 input-4 output switching element implemented by a custom VLSI chip. Eight custom VLSI switch chips were packaged on a single printed circuit board to implement a 16 input-16 output switch.

There was a path through the switch network from each Processor Node to every other Processor Node. Switch operation was similar to that of a packet switching network. The switching nodes used packet address bits to route the packet through the switch from source to destination Processor Node. Each node used two bits of the packet address to select one of its 4 output ports. To illustrate how this works, Figure 3.5-4 shows a packet in transit through the 16 input-16 output switch of Figure 3.5-3. To send a message to node #14, node #5 builds a packet containing the address of node #14 (=1110 binary) followed by the message data and sends the packet into the switch. The first switching node strips the two least significant address bits (10) off of the packet and uses them to switch the remainder of the packet out of its port 2 (10). The next switching node strips

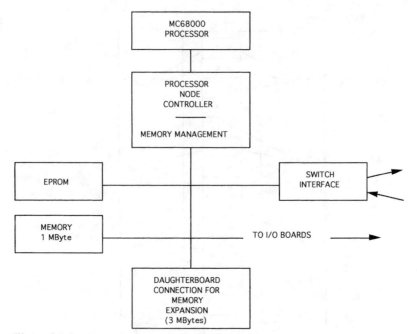

Figure 3.5-2.   Butterfly Processor Node block diagram.

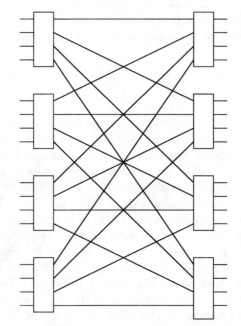

Figure 3-5.3.   A 16 input-16 output Butterfly Switch.

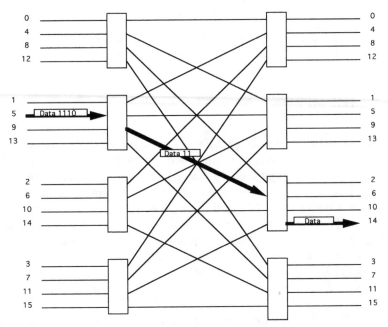

Figure 3.5-4.   A packet in transit through a Butterfly Switch.

off the next two address bits (11) to switch the packet out its port 3 (11) to node #14. Notice that the structure of the switch network ensures that packets with address 1110 will be routed with the same number of steps, to node #14 regardless of the Processor Node sending them.

The Butterfly Parallel Processor was programmed exclusively in a high-level programming language. Most Butterfly software was written in C. Other languages installed on the Butterfly system included Lisp and Fortran. Application programs ran under the Butterfly Chrysalis Operating System.

Programs for the Butterfly system were written using a cross-compiler and other software development tools on a front-end machine. These C language development programs ran under UNIX™ on a DEC VAX or Sun Microsystems workstation. This allowed the use of the rich set of available UNIX software tools, such as the source code control system and the "make" utility. The Butterfly system was connected to the front-end by an Ethernet connection or a serial line. The typical development cycle consisted of editing, compiling, and linking a program on the UNIX front-end, and downloading, running, and debugging the program on the Butterfly system. A source language debugger for the C language was available that ran on the front-end allowing cross-network debugging of programs running on the Butterfly system.

REFERENCE

1. Butterfly Parallel Processor Overview, BB&N Report No. 6148, 6 March
   1986, BB&N Advanced Computers, Inc., Cambridge, MA.

3.6   THE MPP

The Massively Parallel Processor (MPP), an early version of the advanced computer architecture termed single-instruction stream multiple-data stream (SIMD),
promised enormous computational power at lower cost than other existing architectures of its day [1, 2]. Its computational element, the array unit, consisted of a
128×128 array of small 1-bit processors, each containing 1024 bits of local
memory, and having nearest neighbor connectivity. A secondary storage unit, the
staging memory, held 32 Mbytes of data and connected to the array memory via
an 80 Mbyte/s data path. An array control unit broadcast control signals to all
processors in the array unit. The MPP was a back-end processor for a VAX-
11/780 host, which supported its program development and data needs (Figure
3.6-1).

 The MPP was built for the Goddard Space Flight Center by Goodyear Aerospace Corporation and delivered in May 1983. At that time, the construction of a

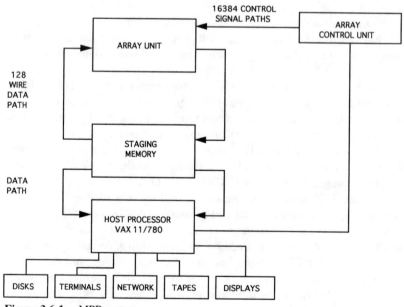

Figure 3.6-1.   MPP system.

digital processor using the very high degree of parallelism embodied in the MPP had not been previously attempted.

After its delivery to Goddard, an extensive language system and a unique operating system were implemented for the MPP. The initial high-level language implemented in 1983 was Parallel Pascal. This language was designed to be independent of computer architecture, thus allowing portability of applications programs between diverse parallel computers having Parallel Pascal compilers. Experience gained in the development and use of this approach revealed that the MPP's 128 × 128 square grid architecture could not easily be hidden from the programmer by using then-current compiler writing technology. A modified language, MPP Pascal, was then implemented that was architecturally dependent, possessing important semantic features that allowed the programmer to make very efficient use of the hardware's capabilities. The MPP Pascal compiler was capable of producing highly optimized code and was flexible enough to allow easy modification. The MPP was also programmable in assembly language.

The MPP operating system provided interactive debugging aids in addition to support for running applications code. The software that performed these tasks was shared by all MPP users, greatly reducing the demand on the host's main memory. The debugging aids included performance monitoring, error reporting or MPP hardware detected faults, breakpointing, single-step, and status display. A first-come first-served queue was the central arbiter controlling user access to the MPP.

All MPP applications programs had to be prepared in two parts. One part ran in the MPP control unit; the other part ran on the host. They were linked together through a message passing system. A master/slave control relationship existed between the MPP and the host. The host resident program was the highest level of control. This program interacted with the user and started the MPP program. MPP programs, in turn, used the host as an I/O server, directly accessing the host's disk and image analysis terminals through an extensive set of I/O service routines.

A device driver that communicated directly with the MPP hardware ran at the lowest level in the host operating system. This driver was the hub of the entire system, controlling the execution of programs in the MPP, as well as the flow of data throughout the system. The bulk of the operating system interacted directly with this device driver to accomplish tasks in support of a running application, such as initializing the hardware, loading programs, starting and stopping programs, reading and writing data, and delivering messages between running programs in the host and the MPP.

For many applications, having only 1024 bits of memory available to each of the 16,384 MPP processors was a serious constraint. The memory chip technology available in 1980 when the system was designed imposed this limitation. As an alternative to an expensive hardware upgrade, a Bit-Plane I/O software package was developed that treated the staging memory as individual bit planes. A system was implemented that provided each processor with 16 K-bits of virtual array memory. The penalty was an increase in memory access time from 0.1 to 25 μs per bit plane. However, many applications benefited from this virtual memory, as they effectively overlapped computation with data transfer. In addition to Bit-Plane I/O, another system, SMM I/O, gave the user access to the powerful data reformatting capabilities of the staging memory.

The Array Unit (the 128 × 128 processing element (PE) array) supplied the MPP's computational power. Each PE had a local 1024-bit random access memory and was connected to its four nearest neighbors—north, south, east, and west. Opposite array edges could be connected together to form either a plane, a horizontal cylinder, a vertical cylinder, or a torus. Arithmetic and logic in each PE were performed in bit-serial manner. All operands were located in the 1024-bit local memory. The cycle time was 100 ns. Table 3-3 shows the raw computing speeds for selected arithmetic operations. The data-bus states of all 16,384 PEs were combined in a tree of inclusive-OR logic elements whose single wire output was used in the Array Control Unit for operations such as finding the maximum or minimum value in parallel of an array.

A single PE is shown functionally in Figure 3.6-2. The P-register, together with its input logic, performed all Boolean logic functions on two variables and could also receive data from the P register in any one of its four nearest neighbors. The A, B, and C registers, the shift register, and associated logic formed an arithmetic unit. The G register controlled masking of arithmetic, logic, and routing operations. (Unmasked operations were performed in all PEs. Masked operations were only performed in those PEs where the G register was set.) The S register

TABLE 3-3

SPEED OF TYPICAL MPP OPERATIONS	
Operation	**Execution Speed, MOPS***
Addition of arrays	
8-bit integers (9-bit sum)	6553
12-bit integers (13-bit sum)	4428
32-bit floating-point numbers	430
Multiplication of arrays	
8-bit integers (16-bit product)	1861
12-bit integers (24-bit product)	910
32-bit floating point numbers	216
Multiplication of array by scalar	
8-bit integers (16-bit product)	2340
12-bit integers (24-bit product)	1260
32-bit floating-point numbers	373

*Million operands per second.

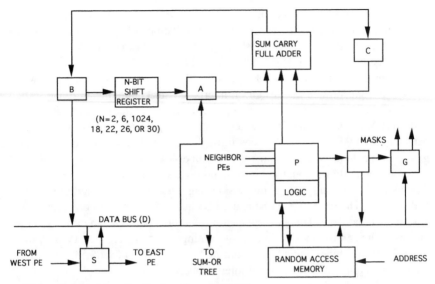

Figure 3.6-2. Functional units of one MPP PE.

was used to shift data to and from the staging memory without disturbing PE operations. A custom integrated circuit (IC) held eight PEs, exclusive of the 1024 bits of random access memory that were on a separate IC chip. This chip, containing a two-row by four-column array of PEs, used high-speed complementary MOSFET (HCMOS) technology. A 1-bit wide bidirectional data-bus connected the memory and the internal components of the PE.

References

1. Frontiers of Massively Parallel Scientific Computation, NASA Conf. Publ. 2478, NASA/Goddard Spaceflight Center, Greenbelt, MD, 1986.

2. Technical Summary: The Massively Parallel Processor, Frontiers '86, p. 293

3.7   THE CONNECTION MACHINE-2

The Connection Machine-2 (CM) was a product of Thinking Machines Corporation (TMC) (Cambridge, MA) [1, 2]. The company was founded in June 1983, about the same time as the Japanese government was initiating its Fifth Generation computer project to exploit new computer architecture in its drive for worldwide industrial leadership.

The company's first product, the 1000 MIPS CM-1, used 65,536 (64 K) single-bit processors. The first commercial installation, a 16 K-sized version, was at MRJ, Inc. (a Perkin-Elmer subsidiary), in August 1986. The 2500 MFLOPS CM-2 was introduced in April 1987. The CM-2 had $16\times$ more memory (64 K bits per processor rather than 4 K), a faster clock, and floating-point hardware.

Hillis concentrated his design on interprocessor communication. He settled on a hypercube connectivity. In this scheme, there are 2^N nodes (N integer). Each node has N directly connected nodes (nearest neighbors) and the longest path between any two nodes has N steps through other nodes. In this context, a single node machine is a zero-dimensional cube, a machine with two connected nodes is a one-dimensional cube, four nodes connected as a square is a two-dimensional cube, and in three dimensions eight nodes form the traditional cube, and so forth. The term hypercube is used because 65,536 nodes corresponds to a 16-dimensional cube. This interconnect design turns out to be an excellent compromise between the mesh connectivity arrangement of the Illiac IV and some other SIMD machines and a fully connected design, which for 65,536 processors would be an impractical number of interconnects.

The flow of control was handled entirely by the front end, including storage and execution of the program and all interaction with the user and/or programmer. The data set, for the most part, was stored in the Connection Machine memory. In this way, the entire data set could be operated on in parallel through commands sent to the Connection Machine processor by the front end. The front end could also operate upon data stored in individual processors in the Connection Machine, treating them logically as memory locations in its virtual memory.

The Connection Machine system implemented data parallel programming constructs directly in hardware. The system included 65,536 physical processors, each with its own memory. Parallel data structures were spread across the data processors, with a single element stored in each processor's memory. When parallel data structures had more than 65,536 data elements (the normal case), the hardware operated in virtual processor mode, presenting the user with a larger number of processors, each with a correspondingly smaller memory.

Communication among elements of a parallel data structure was implemented by a high-speed routing network. Processors that hold interrelated data elements stored pointers to one another. When data were needed, they were passed over the routing network to the appropriate processor.

Scalar data were held in a front-end processor. The front end also controlled execution of the overall data parallel program. Program steps that involve parallel data were passed over an interface to the Connection Machine parallel processing unit, where they were broadcast for execution by all the processors at once.

The Connection Machine front end provided the programming environment for the system. Programs could be stored on front-end disks. Network communications links were most effectively implemented on the front end as well.

High-speed transfers between peripheral devices and Connection Machine memory took place through the Connection Machine I/O system. All processors in parallel passed data to and from I/O buffers. The data were then moved between the buffers and the peripheral devices. Connection Machine high-speed peripherals included the DataVault mass storage system and the Connection Machine graphics display system.

The Connection Machine system software was designed to utilize existing programming languages and environments as much as possible. The languages were based on well-known standards and the extensions to support data parallel constructs were minimal so that a new programming style was not required. The CM-2 front-end operating system (either UNIX or Lisp) remained largely unchanged.

The *Lisp and CM-Lisp languages were data parallel dialects of Common Lisp (a version of Lisp that was standardized by ANSI technical committee X3 J13). *Lisp gave programmers fine control over the CM-2 hardware while maintaining the flexibility of Lisp. CM-Lisp was a higher-level language that added small syntactic changes to the language interface and created a very powerful data parallel programming language.

The C* language was a data parallel extension of the C programming language (as described in the draft C standard proposed by ANSI technical committee X3 J11). C* programs can be read and written like serial C programs; the extensions are unobtrusive and easy to learn.

The assembly language of the CM-2 was Paris. This is the target language of the high-level language compilers. This language logically extended the instruction set of the front end and masks the physical implementation of the CM-2 processing unit.

The Connection Machine Model CM-2 was a computing system that provided both development and execution facilities for data parallel programs. Its hardware consisted of a parallel processing unit, from one to four front-end computers, and an I/O system that supported mass storage and graphic display devices (Figure 3.7-1). The user interacted with the front-end computer; all program development and execution took place within the front end. Because the front-end computer ran standard serial software, the user saw a familiar system environment with additional languages and utilities.

The central element in the system was the CM-2 parallel processing unit, which contained

- thousands of data processors;
- an interprocessor communication network;
- one or more sequencers;
- an interface to one or more front-end computers; and
- zero or more I/O controllers and/or framebuffers.

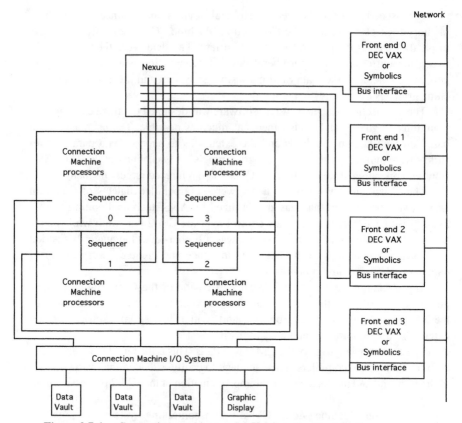

Figure 3.7-1.   Connection machine model CM-2 system organization.

The CM-2 system provided two forms of communication within the parallel processing unit. The more general mechanism was known as the router, which allowed any processor to communicate with any other processor. One may think of the router as allowing every processor to send a message to any other processor, with all messages being sent at the same time. Alternatively, one may think of the router as allowing every processor to access any memory location within the parallel processing unit, with all processors making memory accesses at the same time; in effect, the router allowed the local memories of the data processors to be treated as a single large shared memory. The messages (or accessed fields) were of any length. The throughput of the router depended on the message length and on the pattern of accesses; typical values were 80 million to 250 million 32-bit accesses per second.

The CM-2 parallel processing unit also had a more structured, somewhat faster communication mechanism called the NEWS grid. In the CM-1 and some other fine-grained parallel systems, communication could take place over a fixed two-dimensional grid. The CM-2, however, supported programmable grids with arbitrarily many dimensions. Possible grid configurations for 64K processors included 256×256, 1024×64, $8, \times 8192$, $64 \times 32 \times 32$, $16 \times 16 \times 16 \times 16$, and $8 \times 8 \times 4 \times 8 \times 8 \times 4$. The NEWS grid allowed processors to pass data according to a regular rectangular pattern. For example, in a two-dimensional grid each processor could receive a data item from its neighbor to the east, thereby shifting the grid of data items one position to the left. The advantage of this mechanism over the router was merely that the overhead of explicitly specifying destination addresses was eliminated; for many applications this was a worthwhile optimization.

The parallel processing unit was designed to operate under the programmed control of a front-end computer, which could be a Sun 4, a Symbolics Lisp machine, or a DEC VAX 8000 series computer with a BI bus. The front end provided the program development and execution environment. All Connection Machine programs executed on a front end; during the course of execution the front end issued instructions to the CM-2 parallel processing unit. In effect, the CM-2 parallel processing unit extended the instruction set and I/O capabilities of the front-end computer. The set of instructions that the front end could issue to the parallel processing unit was called Paris. It was designed for convenient use by front-end programs, and included not only such operations as integer arithmetic, floating-point arithmetic, and interprocessor communication, but also such powerful operations as vector summation, matrix multiplication, and sorting.

The data processors did not handle Paris instructions directly. Instead, Paris instructions from the front end were processed by a sequencer in the parallel processing unit. The task of the sequencer was to break down each Paris instruction into a sequence of low-level data processor and memory operations. The sequencer broadcast these low-level operations to the data processors, which executed them at a rate of several million per second.

For every group of 8K data processors there was one I/O channel. (Therefore, a section with 8K processors had one channel; a section with 16K processors had two channels.) To each I/O channel could be connected either one high-resolution graphics display framebuffer module or one general I/O controller supporting an I/O bus, to which several DataVault mass storage devices could be connected. The front end controlled I/O transfers in exactly the same manner that it controlled the data processors, by issuing Paris instructions to the sequencer. The sequencer could then send low-level commands to the I/O channels and interrogate channel status. Data were transferred directly and in parallel between the I/O devices and the data processors, without being funneled through the sequencers.

The Connection Machine Model CM-2 parallel processing unit contained thousands of data processors. Each data processor contained

- an arithmetic logic unit (ALU) and associated latches;
- 64 K bits of bit addressable memory;
- four 1-bit flag registers;
- optional floating-point accelerator;
- router interface;
- NEWS grid interface; and an
- I/O interface.

The data processors were implemented using four chip types. A proprietary custom chip contained the ALU, flag bits, router interface, NEWS grid interface, and I/O interface for 16 data processors, and also contained proportionate pieces of the router and NEWS grid network controllers. The memory consisted of commercial RAM chips. The floating-point accelerator consisted of a custom floating-point interface chip and a floating-point execution chip; one of each was required for every 32 data processors. A fully configured parallel processing unit contained 64 K data processors, and therefore contained 4096 processor chips, 2048 floating-point interface chips, and 2048 floating-point execution chips, and a half a gigabyte of RAM.

A CM-2 ALU consisted of a 3-input, 2-output logic element and associated latches and memory interface. The basic conceptual ALU cycle first read two data bits from memory and one data bit from a flag; the logic element then computed two result bits from the three input bits; finally, one of the two results was stored back into memory and the other result into a flag. One additional feature was that the entire operation was conditional on the value of a third flag; if the flag was zero, then the results for that data processor were not stored after all.

Interprocessor communication was accomplished in the CM-2 parallel unit by special-purpose hardware. Message passing happened in a data parallel fashion; all processors could simultaneously send data into the local memories of other processors, or fetch data from the local memories of other processors into their own. The hardware supported certain message-combining operations; that is, the communication circuitry could be operated in such a way that processors to which multiple messages were sent received the bitwise logical OR of all the messages, or the numerically largest, or the integer sum.

Each CM-2 processor chip contained one router node that served the 16 data processors on the chip. The router nodes on all the processor chips were wired together to form the complete router network. The topology of this network happened to be a boolean n-cube, but this fact was not apparent at the Paris level. (In actuality the 64 K CM-2 was a 12-dimensional cube, with each vertex having 16 fully connected processors.) For a fully configured CM-2 system, the network was a 12-cube connecting 4096 processing chips. Each router node was connected to 12 other router nodes; specifically, router node i (serving data processors 16i through 16i + 15) was connected to router node j if and only if $|i - j| = 2^k$ for some integer k, in which case we say that routers i and j are connected along dimension k.

Each message traveled from one router node to another until it reached the chip containing the destination processor. The router nodes automatically forwarded messages and performed some dynamic load balancing. For example, suppose that processor 117 (which is processor 5 on router node 7, because 117 = $16 \times 7 + 5$) had a message M whose destination was processor 361 (which is processor 9 on router node 22). Since $22 = 7 + 2^4 - 2^0$, this message must traverse dimensions 0 and 4 to reach its destination. In the absence of congestion, router 7 forwarded the message to router 6 ($6 = 7 - 2^0$), which forwarded it to router 22 ($22 = 6 + 2^4$), which delivered the message to the processor 361. On the other hand, if router 7 had another message that needed to use dimension 0, it could choose to send message M along dimension 4 first, to router 23 ($23 = 2 + 2^4$), which then forwarded the message to router 22, which then delivered it.

In addition to the bit-serial data processors described in the preceding, the CM-2 parallel processing unit had an optional floating point accelerator that was closely integrated with the processing unit. There were two possible options for this accelerator: Single Precision or Double Precision. Both options supported IEEE standard floating-point formats and operations. They each increased the rate of floating-point calculations by more than a factor of 20. Taking advantage of this speed required no change in user software.

The Connection Machine I/O structure allowed data to be moved into or out of the parallel processing unit at aggregate peak rates as high as 320 Mbytes/s for a system with multiple I/O controllers. Input/output was done in parallel, with as many as 2 K data processors able to send or receive data at a time. All transfers were parity checked on a byte-by-byte basis.

The specifications in Table 3.7-1 assume a fully configured Connection Machine model CM-2 system with 64 K data processors and eight I/O channels.

TABLE 3.7-1

SINGLE PRECISION FLOATING POINT	
Addition	4000 MFLOPS
Subtraction	4000 MFLOPS
Multiplication	4000 MFLOPS
Division	1500 MFLOPS
4 K × 4 K matrix multiply benchmark	3500 MFLOPS
Dot product	10,000 MFLOPS

Specifications for floating-point performance assume the use of a floating point accelerator.

REFERENCES

1. CM-2 Technical Summary, Thinking Machines Corp., Cambridge, MA.
2. Hord, R. Michael, *Parallel Supercomputing in SIMD Architectures,* CRC Press, Boca Raton, FL, May 1990.

Chapter 4

SIMD

Four of the major single-instruction-multiple-data parallel supercomputers are described: DAP, MasPar, GAPP, and ASPRO. Other SIMD machines now no longer available commercially include the Wavetracer and the Princeton Engine. Research SIMD computers of the past include the GAM and the CLIP4.

Although the SIMD paradigm is no longer the preeminent parallel super-computing scheme, it is anticipated that future architectures will be hetero-geneous with several architectural types of processors embodied in a single machine, and SIMD is expected to be one of those employed. The rationale for this prediction is that no single architecture is the best architecture for all parts of almost any real-world program; programs and algorithms are heterogeneous and so will be the computers on which they are executed. For those parts of a program that lend themselves to SIMD processing, a SIMD subsystem can be expected to be included since the hardware and software costs are comparatively low.

4.1 DAP

The prototype DAP (Distributed Array of Processors) was produced in 1976 by International Computers Limited (ICL) in England. Deliveries of the first genera-tion of DAP computers started in 1980 [1].

In the DAP, the processors (known as "processor elements," or PEs) are arranged in a square matrix: 32 × 32 in the case of the DAP 500 range and 64 × 64 for the DAP 600. Note that the edge size of the matrix is itself a power of 2, the power being expressed in the first digit of the model identifier (i.e., 2^5 for the 500 range and 2^6 for the 600 range).

Each PE is provided with connections to its four nearest neighbors. In ad-dition, a bus system connects all the PEs in each row and all the PEs in each col-umn. These row and column data paths provide rapid data broadcasting or fetching facilities. The nearest neighbor connection scheme gives the high level of connectivity required for many applications (fluid flow, linear algebra, etc.).

Each processor is connected to a minimum of 32 kbits (the architecture allows up to 1 Mbit) of its own local memory. This gives a total memory configuration for the DAP 500 range (1024 processors) of between 32 and 1024 Mbits. The total memory in the DAP 600 range (4096 processors) is between 128 and 4096 Mbits.

In the DAP, the processor array is controlled by a master control unit (MCU). The MCU acts like a conventional central processing unit except that it does not itself execute all the instructions it interprets from its code memory. Many instructions are decoded by the MCU and broadcast to the entire array of processors. Such instructions are executed by all the PEs simultaneously, each operating on the data in its local memory.

The PEs perform their basic operation, which usually involves fetching or storing a memory bit, within a single DAP cycle. On the DAP 510, with a cycle time of 100 ns, there are 10 million cycles per second. The 1024 processors performing logic (Boolean) operations at this rate give a total rate of 1024×10^7 or 10,240 million operations per second (i.e., 10^4 MOPs). A DAP 600 with the same cycle time would increase the final rate by a factor of 4, since the number of processors increases in this ratio. The data rate between memory and processors on the DAP 510 is 10^{10} bits/s, or 1200 Mbytes/s. This bandwidth gives the DAP a very high performance both for computing and for I/O operations.

The DAP is programmed using conventional languages, although to take advantage of the massive parallelism, it is necessary to make extensions to current languages to support data parallel constructs, operations on arrays, for example. Such extensions will be part of the standards of future languages, such as FORTRAN, whose constructs are well suited to the DAP architecture. The current DAP version of FORTRAN is FORTRAN-Plus; it contains a number of extensions and intrinsic routines that make it very easy to take advantage of data parallelism. An assembler language, APAL, is available for anyone who wishes to program the DAP at the bit level for special reasons, such as a requirement for unusual precision or special efficiency.

DAP programs reside in the code memory of the master control unit. The DAP may be accessed through a Sun or VAX workstation, called the DAP host, in order to take advantage of their rich interactive environment and peripheral devices. The user's program can be developed on the host, using a simulator and debugging tool. At run time, the DAP program is initiated and controlled from the host and can communicate with a user program on the host. It depends on the application whether it is better to run a program almost wholly on the DAP or to run only highly data parallel routines on the DAP, leaving the host program predominantly in control.

Although, for many applications, it may be suitable to perform all input/output through the host and use host peripherals, there is a facility for direct I/O to the DAP internal bus using fast data channels at up to 50 Mbytes/s. This can be used to provide real-time video output of the changing data structures of a program during processing. Such I/O rates can be achieved with a negligible effect on the DAP processing speed.

A DAP system has the following main components, as outlined in Figure 4.1-1.

- Master Control Unit (MCU) and Code Memory
- Processor Element (PE) Array and Array Memory
- Host Connection Unit (HCU)
- Fast Data Channel

The Master Control Unit (MCU) is the source of instructions for the DAP. It is a 32-bit central processing unit with many conventional features such as registers, instruction counter, branch instructions, arithmetic unit, and so on. The object code of a DAP program is loaded into the code memory, from which the MCU fetches instructions and interprets them. Some instructions will be executed wholly within the MCU (scalar operations using MCU registers, control instructions, etc.), others will be broadcast to the PE array to be executed by the individual PEs in parallel.

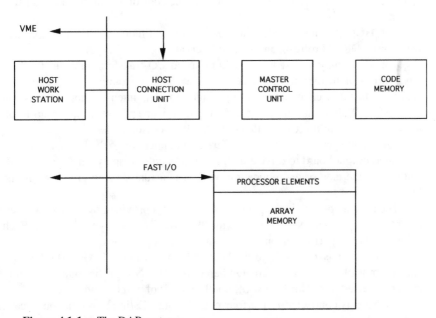

Figure 4.1-1. The DAP system.

Most instructions take place in one cycle. On the DAP 500, the cycle speed is 10 MHz.

The DAP is designed for attachment to a host workstation, which is used for program development, debugging, loading, initiating, and controlling DAP programs. The host connection unit (HCU) serves as the communications gateway between the DAP and the host. The HCU incorporates a Motorola 68020 32-bit microprocessor, a SCSI port, a VME-bus interface, and two RS232 serial ports. Interfacing to VAX computers is via a DR11W or DRB32 interface card.

The DAP HCU provides memory protection through two memory address boundary checkers. A 256-kbyte, or 64 K-word, EPROM provides code storage; and a 1-Mbyte, or 256 K-word, parity-protected random access memory supports data and program code storage.

Data transfers between the HCU and the DAP are performed as memory-to-memory transfers across the VME bus. By using this architecture, any VME bus master can access the DAP memory. The HCU also supports a full VME interrupt handler.

The two RS232 communications ports offer maximum transfer rates of 38.4 and 125 kBd, respectively.

A 1-kbyte FIFO buffer interfaces the SCSI controller chip to the 68020 data bus. Data transferred to and from the bus goes through the SCSI FIFO buffer that is 32 bits wide on the 68020 side and 8 bits wide on the SCSI controller chip side. Data read from and written to the registers to the controller chip bypass the FIFO buffer.

The HCU provides a calendar clock and four timers: bus time-out, system watch-dog, DRAM refresh, and system interval.

The HCU may be used for medium speed I/O transfers either to the host workstation or to directly connected devices on, for example, the VME port. However, fast data channels are also provided for data intensive applications. On the DAP 510, data may be input and output concurrently at a rate of up to 50 Mbytes/s; this would use only 4% of the DAP processor cycles.

Among the peripherals that are offered for connection to the fast data channel is a video output board to drive a high resolution color display, enabling "visualization" of the data of an application while it is actually being processed in the DAP.

The parallel processing capability of the DAP is provided by N^2 processor elements arranged in an $N \times N$ matrix (the "PE array" or simply the "array"). Each PE is capable of performing arithmetic and logical operations.

N (the "edge size") is 32 in the DAP 500 and 64 in the DAP 600. The rows and columns of the PE matrix are numbered from 0 to $N - 1$. The edges of the matrix are referred to as North (row 0), South, West (column 0), and East.

Each PE is connected to the four neighboring PEs in the north, south, east, and west directions. Using these connections, data can be propagated from a register of each PE into the corresponding register of a neighboring PE in any of the four directions. Processors at an edge are connected to those at the opposite edge, thus allowing shifts to wrap around if desired. Also, the PEs in each row and in each column are connected on row and column highways to broadcast data in and out of all the processors simultaneously.

Figure 4.1-2 is a simplified diagram of one processor element. This diagram is intended only to show the main functional components and their interconnection. Each PE has three 1-bit registers, denoted A, Q, and C. They are used for a number of purposes, but it is convenient to think of the Q register as being an accumulator, the C register as a carry register, and the A register as being for "activity control"—this means it can inhibit memory write operations in certain instructions. All of the bits of a particular register over all the PEs are known as a "register plane."

The Q and C registers are input to an adder, but these inputs can be disabled (i.e., treated as false). In this way, the adder can take data from either, none, or both of the Q and C registers. The third adder input is selected by a multiplexer and can come from the PE memory, the outputs of the Q or A register, data broadcast by the MCU, or the carry output of a neighboring PE. The A register also receives its input from this multiplexer and can either be written directly or AND-ed (masked) with the existing A register contents. For some purposes, the output of the input multiplexer can be inverted. PE outputs can be written to memory, and in some instructions this writing is conditional on the value in the A register.

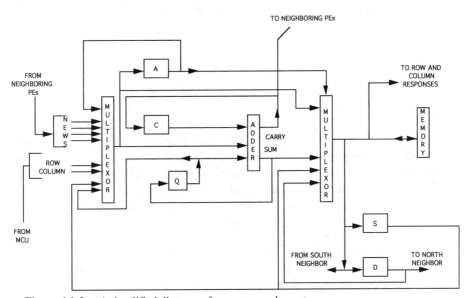

Figure 4.1-2. A simplified diagram of a processor element.

The D register shown in Figure 4.1-2 does not appear explicitly in the machine-code programmer's model of the PE, but is used for input or output for the fast interface unit. Once data is loaded into the plan of D registers, it can be clocked out asynchronously without interrupting the processing.

Certain instructions both read from and write to the memory; for these instructions a register is needed somewhere in the path of the data flow. The details are implementation dependent, but an example of such a register is shown as S in Figure 4.1-2; this register does not appear in the programmer's model.

Each of the PEs has a local memory whose size depends on the version of the DAP but will be between 32 K (32,768) bits and 1 Mbit. The sum total of this PE memory is referred to as the DAP array memory.

Two mappings particularly well suited to the DAP structure follow.

- Vector (or horizontal) mode, in which successive bits of a word are mapped onto successive bits of a single row of a store plane
- Matrix (or vertical) mode, in which successive bits of a word are mapped onto the same bit position in successive store planes

The master control unit (MCU) performs the following functions:

- instruction fetching, decoding, and address generating;
- executing certain instructions, and broadcasting other instructions to the PE matrix for simultaneous execution by all PEs;
- transmitting data between the array memory of the PE array and the MCU registers;
- providing hardware support for DO loop instructions;
- supporting data transfer between the DAP and the host filestore or attached peripheral devices.

The object code of a DAP program resides in the code memory known as the "codestore." Instructions are all 32 bits long. The size of the codestore depends on the model of the DAP, but it is between 128 K words (512 kbytes) and 1 M word (4 Mbytes). A schematic diagram of the MCU is shown in Figure 4.1-3.

The MCU has a number of 32-bit general-purpose registers that are visible to the machine language level programmer.

Registers can be loaded in various ways from the array memory, or a register's contents can be supplied as data to the memory or PEs. Logical and arithmetic operations are available to operate on the registers. Programs can test the contents of the registers and skip on certain conditions, or use them to hold link

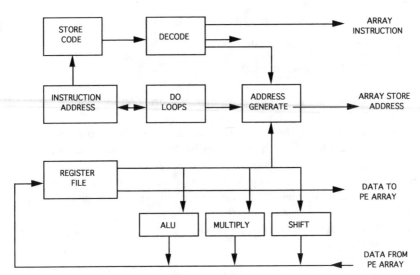

Figure 4.1-3.   Schematic diagram of the master control unit.

values for subroutine entry and exit. A register's contents can modify addresses or values; a register being used in this way is referred to as a modifier register. In all, there are 15 registers available to the machine-code programmer.

- M0 to M13 are general purpose registers, which can hold link values or data, and are operated upon by MCU arithmetic or logical functions. The contents of these registers can also be transferred to or from the array.

- M1 to M7 can always be used as modifiers. Register M0 is not generally available as a modifier, since value 0 in the instruction modifier field is interpreted as no modification.

- ME is the edge register, and is matched in size to the array edge dimension. In the DAP 500, since the edge size is 32 bits, ME is the same size as the other registers. The ME register is regarded as part of the array and can be used as the source or destination of data transferred to or from the array; it cannot in general take part in MCU arithmetic or logical operations or act as a modifier register.

In addition to the MCU registers, there are two 1-bit flags, the C-flag (carry) and V-flag (overflow) written to appropriately by MCU scalar arithmetic instructions on 32-bit signed or unsigned values.

There are three machine states.

1. 0: Nonprivileged, interruptible (this is user mode)

2. 1: Privileged, interruptible

3. 2: Privileged, noninterruptible

The machine state affects the generation of instruction addresses and data addresses and affects the legality of certain instructions.

When in user mode, both codestore and array memory addresses are partitioned by "datum/limit" registers. Actual hardware addresses are obtained by adding the appropriate datum register, and a check is made that the limit is not exceeded. This permits protection of supervisor space from user space and also allows secure multiprogramming between user programs. User programs run in state 0 and deal only with addresses that are relative to array or code datum.

Instructions generate array memory addresses that, in the most general case, specify the following.

a 32-bit word

within a row or a column

within a store plane

Twenty (20) bits are used for the store plane address in the DAP (thereby giving the architectural limit of 1 Mbit memory within each PE). This part of the effective address is known as the ADDR field.

Five (5) or 6 bits are used for the row or column address within a bit plane, depending on the model of the DAP: the DAP 500 (32 × 32) or DAP 600 (64 × 64). This is known as the INT field.

Zero or 1 bit is used to specify a word within a given row/column, depending on the DAP model. This is known as the W field. (In the DAP 500, a word is the whole row or column and no separate word field is required.) Thus the full (32-bit) effective address takes the following form.

```
.... ....A AAAA AAAA AAAA AAAA AAA1 1111 DAP 500
.... .AAA AAAA AAAA AAAA AAAA A111 111W DAP 600
```

where A = ADDR field bits, 1 = INT field bits, and W = W field bits. Some instructions operate on whole store planes; in these cases the INT field of the effective address will be ignored or used for some other purpose. Some instructions operate on rows or columns rather than words, in which case the W field is not relevant to the array memory address.

In the most general case, an address is generated by a combination of four quantities.

1. An address field in the instruction itself—this normally allows only 8 bits for the ADDR field.

2. Modification by an MCU register specified in the instruction.

3. Addition of a DO loop step on INT and ADDR fields.

4. Datum register.

A hardware DO instruction is used to invoke a "DO loop," a sequence of instructions executed a specified number of times unless a premature exit is taken. These loops are used for operations such as adding successive bits of the operand. Provision of DO as a hardware feature means that there are no loop control overheads for these cases.

A feature of the DO loop is the facility for instructions operating in it to access, on successive iterations of the loop, successive bit planes, rows, columns or words of memory, or bits in an MCU register. This process is referred to as "address stepping." It is implemented by adding the DO loop iteration number to, or subtracting it from, addresses or values to give the addresses or values that are "effective" for that particular execution of the instruction. This is in addition to any register address modification. The iteration number used for address stepping is zero in the first pass of the loop and is incremented at the end of each pass of the loop.

Instructions that result in information being propagated through the nearest neighbor connections of the PEs—namely, the vector add and plane shift instructions—generate "direction, geometry, and count fields." These can be specified in the instruction or picked up in a modifier register.

The "direction" is N, S, E, or W. In the case of shifts, it gives the direction of the shift. In the case of vector adds, it specifies whether rows (E, W) or columns (N, S) are being added, as well as the direction of the propagation of the carry bits (i.e., which end is regarded as the most significant). The geometry specifies the behavior at the edges on shifts or carries: "cyclic geometry" implies wraparounds, "plane geometry" implies no wraparounds, with zeros being introduced at the edge. Plane and cyclic geometry can be independently specified for N/S and E/W. The count field can give the degree of shifting and the degree to which carries are propagated.

There are 16 groups of instructions. Of these, the first 14 groups contain array instructions, acted on by PEs. The operands of these refer typically to PE registers and a store plane, and the instruction will be broadcast to each PE to operate on its own bits within these planes. Some of the instructions move data between MCU registers and either 32-bit words or row/columns within the array. Note that in the latter case, fetching a row/column to an MCU register (rather than the edge register) will fetch only the least significant 32 bits; conversely, a memory write operation will expand data in the MCU register to the edge size by zeros at the top end. The edge register is, in fact, provided to assist in the writing of machine code that is compatible across different DAP edge sizes.

The other two groups of instructions (groups E and F) contain MCU scalar instructions performing logical, arithmetic, and control operations. Note that status and control registers are memory-mapped, and there is no need for special instructions to perform, for example, I/O.

Medium speed I/O to the DAP can be achieved through the host or through the VME bus interface. However, fast data channels closely coupled to the DAP array are provided for data intensive applications. The DAP 500 provides transfer rates of up to 50 Mbytes/s both in and out.

REFERENCE

1. DAP Series Technical Overview, Active Memory Technology, Irvine, CA, approx. 1987.

4.2   MASPAR MP-1 AND MP-2: IMPLEMENTATION OF THE NAS PARALLEL BENCHMARKS SIMULATED APPLICATIONS

4.2.1   Introduction

The Numerical Aerodynamic Simulation (NAS) program is a large-scale effort to advance the state of computational aerodynamics. The program, which is based at NASA Ames Research Center, specifically targets the goal of simulating an entire aerospace vehicle system in 1 to several hours, by the year 2000. This "Grand Challenge" problem will require computing systems having a thousand times the performance of current supercomputers. This aggressive objective leads naturally to interest in parallel processing.

Performance evaluation of parallel systems has traditionally been limited to popular kernels such as LINPACK and Livermore Loops. These are clearly inadequate for gauging the capability of a parallel machine for any given aerospace-related problem. On the other hand, utilizing a full application for benchmarking is often time prohibitive and the porting effort to very distinct architectures would limit the availability of performance results across an ever increasing variety of machines. The NAS group determined that constraints required to produce a useful benchmark suite for parallel machines suggested a set of "paper and pencil" tests, where the problem is described at a very high level. In this way the implementor is free to determine the data structures, algorithms, processor allocation, and memory usage to afford optimal porting to a specified target machine.

With these design objectives in place, the NAS group created a set of eight benchmark programs. These fall into two broad categories: five relatively simple kernels and three simulated CFD applications. The kernels measure performance on specific functions frequently encountered in computational aerodynamics and other areas of scientific/engineering computing. Specifically they include, an embarrassingly parallel computation of pseudo random numbers, a 3D FFT PDE solver, a multigrid solver, a conjugate gradient solver, and an integer sort. The simulated applications are all three-dimensional in nature and include an approximate factorization scheme solved via ADI called APPBT, a diagonalized approximate factorization scheme also solved via ADI called APPSP, and a representation of a newer class of CFD algorithms solved using Symmetric Successive Over-Relaxation called APPLU. The same governing equations (1) are used in each, whereas the different methods are used to obtain a solution. The benchmarks can be scaled in size as machine performance increases over time.

This section describes the implementation of the three simulated applications on the MasPar series of high performance, cost effective, massively parallel computers. After an introduction to the MasPar system architecture, implementation details, performance, and cost performance are highlighted. The motivation and mathematics behind these simulated applications are described by Baily et al. [1] and are not addressed in this section, rather a description of the mechanics of the computation is attempted.

The MasPar Computer Corporation MP-1 and MP-2 are fine grain massively parallel Single-Instruction, Multiple-Datastream (SIMD) computers consisting of from 1 to 16 K processing elements [2, 3]. All of the computation presented in this work was performed on machine having 4K processors. The architecture consists of a workstation front end, an array control unit, and the array of processing elements. Architecturally, the MP-1 and MP-2 are the same; however, implementation of this architecture differs substantially.

The Array Control Unit (ACU) is a 12-MIPS scalar processor with a RISC-style instruction set and a demand-paged instruction memory. The ACU fetches and decodes MP instructions, computes addresses and scalar data values, issues control signals to the PE array, and monitors the status of the PE array.

Processing elements are custom RISC style processors using a basic load/store style instruction set design and are clocked at 80 nsec. The MP-1 utilizes a 4-bit processor, whereas the MP-2 uses a 32-bit processor. Each chip contains 32 of these individual PEs. There are 64 Kbytes of off chip memory for each PE. Significant features of the PE include hardware support for memory and computation overlap, hardware supported indirect addressing, and a balanced memory bandwidth. Aggregate peak floating point performance for a 16 K processor machine is 1.2 GFLOPS for the MP-1 and 6.4 GFLOPS for the MP-2.

The architecture allows interprocessor communication to proceed in two ways. To perform random, or long-distance communication, a three-stage cross-bar, circuit switched router may be employed. This general purpose network provides a bandwidth in excess of 1 GigaByte/sec in a 16 K processor system. For more regular communication patterns, the machine may be viewed as a two-dimensional grid with each PE having an eight-way connection to its neighbors with toroidal wraparound. This second network, termed X-Net, provides an aggregate bandwidth exceeding 20 GigaBytes/sec in a 16 K processor system.

All of the applications were programmed using MPFORTRAN, MasPar's FORTRAN 90 implementation [4]. NAS guidelines for implementing the benchmarks require all calculation to be performed in double precision floating point. The three-dimensional grid is of size N^3, where N is the edge length along each spatial dimension. For all runs, the suggested grid size is N = 64.

4.2.2   APPBT: Approximate Factorization (Beam-Warming)

The APPBT simulated application utilizes an ADI scheme to solve an approximate factorization of the governing equations. The factorization, coupled with the use of the ADI technique, leads to a requirement for the simultaneous solution of N^2 block tridiagonal systems of length N (block size = 5) three times per time step. Each of these three ADI sweeps treats a different spatial dimension implicitly.

In mapping this problem to the MasPar MP-x series of computers, it is first important to note that the topology of processing elements is a two-dimensional grid. It is logical to assign two of the three spatial dimensions of the problem grid such that neighboring points are on adjacent processors. Nearest neighbor communication that arises naturally from difference style computation across these problem dimensions is handled efficiently by the aforementioned X-net network. The third spatial dimension is mapped into the memory of processors.

Most of the required computation in solving this problem occurs in the following segments of the calculation.

- Forming the explicit right-hand side (once per time step).
- Forming the required flux Jacobians (once per ADI sweep).
- Solving the N^2 length N block tridiagonal systems (once per ADI sweep).

Parallelization of the explicit right-hand side calculation and calculation of the flux Jacobians is straightforward as there are no troubling data dependencies. Solving the systems of block tridiagonal systems can be addressed in a number of ways, which we shall now examine.

With the data placement described in the preceding, one of the ADI sweeps, specifically the one that treats the problem axis mapped to memory implicitly, results in block tridiagonal systems that can be handled with the best serial algorithm. Treatment of the other two spatial dimensions is not as obvious. It is tempting to apply a parallel algorithm such as Parallel Cyclic Reduction [5, 6], but the performance penalty is significant as compared to the sweep using the serial algorithm.

It is more appropriate to consider remapping data between sweeps in such a way that the best serial algorithm can be used for each. In this approach, the problem state variables and the explicit right hand side vectors in the implicitly treated direction are mapped into memory for every sweep. This permits the formation of the required Jacobians and the subsequent solution of the block tridiagonal systems to be done in a serial fashion on each processor. The only significant amount of communication occurs when remapping the data for each sweep; however, this only amounts to less than 5% of the total runtime in both of the MasPar implementations.

4.2.3   APPSP: Diagonal Form of Approximate Factorization

Here the ADI technique, coupled with the a diagonalization of the approximate factorization leads to $5N^2$ scalar pentadiagonal systems for each sweep direction. The exact same technique as described for the previous application can be used where the only difference is the ratio of communication to computation increases owing to the simplification of the equation sets to be solved. Remapping communication still only accounts for less than 12% of the total calculation time, far less than the additional cost if the data were not remapped and the systems of equations were solved across processors.

4.2.4   APPLU: Symmetric Successive Over-Relaxation

In this simulated application, strict data dependencies determine the mechanics of the calculation. The grid points are distributed to the machine such that the x and y grid directions are mapped to the processor grid and the z dimension are mapped into memory. Two sweeps through the grid are required to advance the solution through a single time step. The first sweep starts at grid point (1, 1, 1) and proceeds plane by plane, each plane having the surface normal (1, 1, 1), until grid point (N, N, N) is reached. All the points on the plane normal to the sweep direction vector may be processed at once, as they only require updated information available from the adjacent plane already processed. The second sweep is exactly opposite, starting at (N, N, N) and proceeding in the (–1, –1, –1) direction until reaching (1, 1, 1).

This data dependency cannot be avoided in this scheme, leading to a situation where on average one-third of the processors are active during the sweeps. The time-consuming steps of the calculation are similar to those of APPBT, namely calculating the explicit right-hand side (once per time step), calculating flux Jacobians (once per sweep), and finally performing the SSOR sweeps (two sweeps per time step). There are no difficult data dependencies in the calculation of the flux Jacobians so these may be precomputed before each sweep. This leads to significant savings as they can be calculated one $x - y$ plane at a time thereby keeping all the processors active. If flux Jacobians are computed as needed during the sweep, we would be limited to using one third the processors on average for this aspect of the calculation as well.

In this study, all performance results were obtained on 4096 processor MP machines. This is most appropriate because the currently published NAS parallel application benchmark results utilize a 64^3 grid and the 4 K MasPar machine consists of a 64^2 grid of processors. Performance results are presented in Tables 4-1 to 4-3 for both the MasPar MP-1 and MP-2 along with the Cray C90 16 processor machine, a single processor Cray Y/MP, an eight processor Cray Y/MP 8, a 128-node iPSC/860, and finally a 32 K node CM-2. Results for non MasPar computer were obtained from reference [7]. Approximate list prices for these various machines, used to calculate price-performance, are provided in Table 4-4.

TABLE 4-1

PERFORMANCE AND PRICE-PERFORMANCE FOR APPBT ON VARIOUS HIGH-PERFORMANCE COMPUTER SYSTEMS*				
Computer (nproc)	MFLOPS (dbl prec)	% Peak	T_{iter} (sec)	Price-Performance (Y/MP = 1)
MP-2104 (4 K)	208	34.7	4.35	4.45
MP-1104 (4 K)	75	54.3	12.10	2.67
C90 (16)	6386	39.9	0.14	2.01
Y/MP (1)	229	68.7	3.96	1.0
Y/MP (8)	1590	59.7	0.57	0.83
iPSC/860 (128)	430	5.6	2.11	1.20
CM-2 (32 K)	95	1.0	9.57	0.23

*MFLOPS is measured for double precision calculation. Price performance is normalized to the price performance of a single processor Cray Y/MP. The number of processors is given in brackets after each computer model name.

TABLE 4-2

PERFORMANCE AND PRICE-PERFORMANCE FOR APPSP ON VARIOUS HIGH-PERFORMANCE COMPUTER SYSTEMS				
Computer (nproc)	MFLOPS (dbl prec)	% Peak	T_{iter} (sec)	Price-Performance (Y/MP = 1)
MP-2104 (4 K)	153	25.5	1.67	3.46
MP-1104 (4 K)	58	41.7	4.43	2.17
C90 (16)	7810	48.8	0.03	2.60
Y/MP (1)	216	64.9	1.18	1.0
Y/MP (8)	1579	59.3	0.16	0.87
iPSC/860 (128)	206	2.7	1.24	0.61
CM-2 (32 K)	94	0.9	2.7	0.24

The remapping method of implementing both APPBT and APPSP is applicable to many three-dimensional problems. For example, 3D FFT calculations can be performed in three highly parallel sweeps by remapping the data between each. When appropriate, it usually enables the use of more efficient algorithms at the cost of doing the remapping. This cost is minimal on the MasPar architecture.

TABLE 4-3

PERFORMANCE AND PRICE-PERFORMANCE FOR APPLU ON VARIOUS HIGH-PERFORMANCE COMPUTER SYSTEMS				
Computer (nproc)	MFLOPS (dbl prec)	% Peak	T_{iter} (sec)	Price-Performance (Y/MP = 1)
MP-2104 (4 K)	98	16.4	2.63	2.48
MP-1104 (4 K)	33	23.9	7.83	1.39
C90 (16)	3665	22.9	0.07	1.37
Y/MP (1)	194	58.1	1.33	1.0
Y/MP (8)	1304	48.9	0.20	0.81
iPSC/860 (128)	146	1.9	1.77	0.48
CM-2 (32 K)	76	0.8	3.4	0.32

TABLE 4-4

APPROXIMATE U.S. LIST PRICES FOR VARIOUS HIGH-PERFORMANCE COMPUTER SYSTEMS	
Computer	List Price (U.S.D.)
MP-2104 (4 K)	449,500
MP-1104 (4 K)	270,000
C90 (16)	30,500,000
Y/MP (1)	2,200,000
Y/MP (8)	18,400,000
iPSC/860 (128)	3,450,000
CM-2 (32 K)	4,000,000

REFERENCES

1. Baily, D., et al., The NAS Parallel Benchmarks, Report RJR-91-002 Revision 2, August 1991.

2. Blank, T., The MasPar MP-1 Architecture, *Proceedings of COMPCON Spring 90—The 35th IEEE Computer Society International Conference*, San Francisco, 1990.

3. Nickolls, J. R., The Design of the MasPar MP-1: A Cost Effective Massively Parallel Computer, *Proceedings of COMPCON Spring 90—The 35th IEEE Computer Society International Conference*, San Francisco, 1990.

4. McDonald, Jeffrey, The MasPar MP-1 and MP-2 Implementation of the NAS Parallel Benchmarks Simulated Applications, MasPar Computer Corp., Sunnyvale, CA, presented to Supercomputing '92, Minneapolis.

5. Johnsson, S. L., Solving Tridiagonal Systems on Ensemble Architectures, *SIAM Journal of Scientific and Statistical Computing*, Vol. 8, No. 3, May 1987.

6. Hockney, R. W. and Jesshope, C. R., *Parallel Computers 2*, Adam Hilger, Philadelphia, 1988.

7. Baily, D., Barszcz, E., Dagum, L., and Simon, H., "NAS Parallel Benchmark Results," RNR-92-002, presented at Supercomputing '92, November 16–20, 1992.

4.3   GAPP

The GAPP concept was first implemented as a medium scale integration (MSI) breadboard in 1982. This first system emulated a 6 × 12-cell array using programmable logic and discrete memory components. The development of GAPP technology continued into 1983 with the commitment to develop a GAPP-based custom integrated circuit. The first design resulted in 3 × 6-cell chips. Prior to the completion of these parts, known as GAPP I, Martin Marietta began to improve the design of the basic cell toward higher cell density per chip (6 × 12 cells or 72 cells per chip). The first of these new chips, GAPP II, was delivered to Martin Marietta in late 1984.

During chip design and development, Martin Marietta designed and built a prototype system. This system was designed to perform real-time (30 frames per second) video processing. The video source of primary interest and focus was a forward looking infrared (FLIR) sensor. This application was chosen because the FLIR is a major product line of Martin Marietta Electronics and Missiles Group in Orlando.

The largest GAPP system's main array contains 82,944 processing elements. This is probably the largest array of processors ever constructed [1].

Martin Marietta intentionally designed the GAPP cell as simple as possible. Nothing should be included in the cell that is not involved in the computation clock cycle. This requirement keeps the cell structure small, allowing a large number of cells per chip. The cell consists of six active components: four 1-bit registers, a 1-bit full adder/subtractor (FAS), and 128 bits of memory. Additionally, multiplexers and data paths permit the movement of signals within the cell [2–4].

Three of the 1-bit registers, labeled North-South (NS), East-West (EW), and carry borrow (C) are connected to the inputs of the 1-bit FAS. Additionally, the NS register output is connected as an alternate input to the NS registers in the cells that exist geometrically to the north and south of this cell. Likewise, the EW register is connected as an alternate input to the EW registers in the cells that exist geometrically to the east and west of this cell. This is the nearest neighbor orthogonal connection of the fine grid array of GAPP cells. The contents of the C register are not available outside the cell in which it exists without passing through some other register.

The 1-bit full adder/subtractor (FAS) is the computational element of the cell. It implements a truth table.

The three 1-bit inputs come from the three previously mentioned registers, NS, EW, and C. On every clock cycle the FAS automatically produces the result prescribed by the truth table. This truth table allows the construction of arithmetic and logical results, in a bit serial fashion, that are completely general. In principle, each cell can perform all arithmetic and logical operations with this element. The output labels represent, respectively, sum, carry, and borrow (SM, CY, BW).

The memory array in each cell is organized as a 1 × 128-bit static RAM. When a 7-bit address is supplied, along with read-write signals, 1 bit of data may be read from or written into every cell's addressed memory location.

Each cell requires 20 bits of control/address information defining the activity required of the cell. The control section consists of 13 signals (the other seven are associated with the RAM addressing) that primarily select data paths: one associated with each portion of the cell that can store data. Thus, the RAM and registers NS, EW, CM, and C can be manipulated in parallel. Additionally, each cell must receive a clock signal. All changes of state within the cell occur synchronously with the clock.

The control, address, and clock signals are common among all cells on the chip. Thus, every cell performs exactly as its neighboring cell. The only difference between activities are a function of the data content within each cell's registers and RAM. These data differences are crucial because a cell or group of cells, through the proper use of algorithms, can appear to be "turned off." The cell's ability to perform logical operations makes individual cell operations practical, even in a SIMD control strategy.

On a single GAPP chip, there are six sets of 12-bit NS shift registers and 12 sets of 6-bit EW shift registers. Since every processing element contains one NS register and one EW register, then these groups of NS and EW shift register form a geometric orthogonal arrangement across groups of cells. The contents of the EW registers to be transferred may use the NS registers and vice versa.

Similarly, the CM registers are organized as a group of shift registers, geometrically placed in parallel with the NS shift registers. The CM registers are unidirectional (from south to north), allowing images to be input from the south and output to the north.

Each grouping of like-named registers, such as outputs from the FAS and RAM locations at the same address, can be thought of as planes of data. When an instruction is executed, every cell in the chip reacts in exactly the same way. Since each chip contains 72 cells, this has the effect of operating on a 72-bit "word" within the chip for up to five planes (instructions involving NS, EW, C, CM, and RAM) in one clock time. Usually one to three planes are moved at once.

The ends of each of the three groupings of shift registers (CM, NS, and EW) come to the edge of the chip. Both the CM and NS groups exit at the north and south edges, whereas the EW group exists at the east and west edges. Each of these data groups may be thought of as input/output ports to the chip. In that sense, each chip has six ports; two bidirectional 6-bit ports (one at the northern and one at the southern edge for NS); two unidirectional 6-bit ports (one for output at the northern edge of CM and one for input at the southern edge of CM); and two bidirectional 12-bit ports (one at the western and one at the eastern edge for EW). Further, the system designer may choose to provide three simultaneously input (CMS, and E or W, and N or S) and three simultaneously output (CMN, and E or W, and N or S) paths on a given clock cycle. At a 10 MHz clock, each chip has an input/output bandwidth of 60 Mbytes/s, 30 MBytes/second input and 30 Mbytes/s output.

The data signals are deliberately pinned out of the chip package at four mechanical edges, providing relatively easy printed circuit board layout.

A cell requires 25 clocks to perform an 8-bit add ($3n + 1$, where n is the number of bits in each operand to be added). Each chip can be clocked at 10 MHz frequency. At this 100 ns rate, each cell can perform 400,000 8-bit adds per second. Each chip contains 72 cells each performing their add; thus, the chip throughout is equivalent to 28.8 million 8-bit adds per second. The 8-bit add executes in 2.5 μs. As an example of an elementary image processing operation, a 3 × 3 neighborhood Sobel operator takes 54.6 μs or 18,315 Sobels per second per cell.

Each chip contains 84 pins for power, ground, clock, control, address, and data exchange. Each chip occupies about one square inch of board space and dissipates about one-half watt.

4.3.1   GAPP Array

The assembly of an array is simple; each chip is connected to its logical neighbor (east connects to west and north to south). Clocks and control are distributed to every chip in the array. Practical limits exist and most are imposed by the choice of board housing, backplanes, bus standards, or system architecture. In the current design, arrays are modularized as 48 × 132 (6336) cells on a single 9 U board (14 3/8 × 16 in).

In standard systems, input of data occurs via the CM south port and output occurs from the CM north port. This arrangement takes advantage of the CM plane, allowing for simultaneous input and output during computation. To obtain simultaneous input and output three conditions must exist. First, a result must be available at the start of the input/output operation. Second, a plane of input data must be available in the external world. Third, the algorithm currently running must require at least N clocks, where N is the size of the GAPP array in the north-south direction. To obtain free input/output, a result is loaded from RAM or registers into the CM plane in one clock cycle. Data in the CM register plane are shifted north one position for each clock. Simultaneously, a row of data are output into an appropriate buffer on the northern edge of CM. On the same clock, a row of data are input into the southern edge of CM. This operation continues for N clocks. During the N clocks, any other operation can occur within the array as long as it does not involve the CM plane. At the end of N clocks, data are written from CM into RAM or registers as directed by the program.

The smallest size array is one chip. The required array size is tailored to the systems problem. In real-time image processing, the major parameters then determine size including input data rate, algorithm length (execution time), and array clock speed. For example, assume a 10-MHz clock speed machine accepting imagery arriving at 12 megapixels per second and that the algorithm requires 50,000 instructions or clocks (the equivalent of 2000 8-bit adds for every pixel in the array).

The algorithm will require 5 ms to execute (50,000/10,000,000). The array must contain at least 60,000 cells (5 ms × 12,000) to maintain real-time rates without missing any data; this equates to about 833 chips. Using the 48 × 132 cell GAPP modules previously mentioned, a system containing 10 modules will suffice (880 GAPP chips or 63,360 cells). The modular design approach can accommodate up to 24 GAPP modules, 2112 chips, or 152,064 cells. At a 10-MHz clock frequency, a 24-module system would exhibit a computational throughput of 60 giga 8-bit adds per second. The largest GAPP systems to date contains 1152 chips or 82,944 cells.

Each processor element contains separate lines that link the cell to its neighbors and the outside world. In addition to the North-South (NS) and East-West (EW) lines that pass data between cells, there are the CM South input (CMS) and CM North output (CMN) (Figure 4.3-1). There is also a complement of 22 external signal lines: seven address lines (A_0 through A_6), 13 control lines (C_0 through C_{12}), one global output (GO), and one clock (CLK).

The chip's overall simplicity is reflected in the layout of a single processor element (Figure 4.3-2). Each of its four latches—CM, NS, EW, and C (referred to as the C register)—accepts data from up to eight possible sources, depending on the setting of the control lines. C_0 and C_1 control the input to the CM latch; C_2 through C_4 govern the input to the NS latch; C_5 through C_7 manage the input to

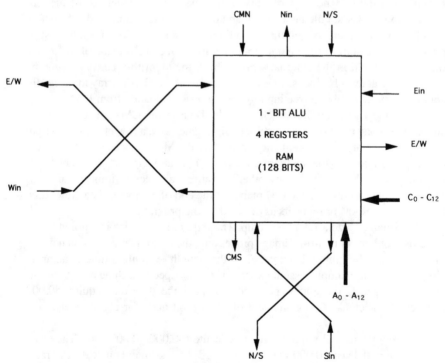

Figure 4.3-1.   GAPP PE links to neighbors.

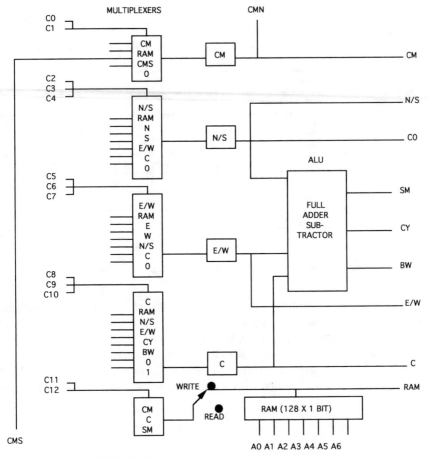

Figure 4.3-2.   GAPP PE diagram.

the EW latch; and C_8 through C_{10}, the input to the C register. Lines C_{11} and C_{12} handle reads to and writes from the 128-bit RAM.

Working from a truth table, the array performs additions and subtractions (Table 4-5). The C, NS, and EW inputs to the multiplexers represent the contents of the C, NS, and EW registers, respectively. The summing output of the single bit ALU, SM, goes directly to the RAM and may also be simultaneously input to any of the four registers. The Carry and Borrow outputs (CY and BW, respectively) are open to the C register. A truth table is used as well to fulfill single- and dual-input logic functions—logical complement, exclusive-OR, exclusive-NOR, logical AND, and logical OR—on data in the NS and EW latches (Table 4-6).

TABLE 4-5

GAPP Truth Table					
Arithmetic Operations					
Input			**Output**		
NS	EW	C	SM	CY	BW
0	0	0	0	0	0
0	1	0	1	0	1
1	0	0	1	0	0
1	1	0	0	1	0
0	0	1	1	0	1
0	1	1	0	1	1
1	0	1	0	1	0
1	1	1	1	1	1

Programming the array is well-suited to tasks such as addition and multiplication. The algorithm for recognizing a 101 pattern demonstrates its ability to take care of both arithmetic and logical operations (Figure 4.3-3).

In the first step, the EW registers of a configuration of 8×5 processor elements are loaded with the test pattern found at RAM location 0 of each cell. Next the entire test pattern is shifted one column to the left. In the third step, those data

TABLE 4-6

GAPP Logic Operations			
Logical Operation	Output	Input	Input Conditions
INV	SM = SM = SM =	$\overline{NS}$ $\overline{EW}$ $\overline{C}$	EW = 0, C = 1 NS = 0, C = 1 NS = 0, EW = 1
AND	C = CY = BW =	NS • EW EW • C $\overline{NS}$ • EW	C = 0 NS = 0 C = 0
OR	CY = BW = BW =	NS + EW $\overline{NS}$ + EW EW + C	C = 1 C = 1 NS = 0
XOR	SM =	NS + EW	C = 0
XOR	SM =	$\overline{NS + EW}$	C = 1

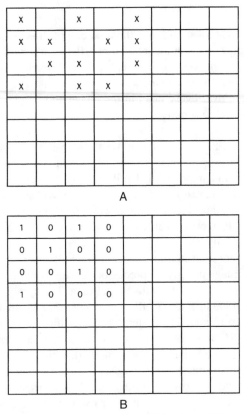

Figure 4.3-3.   GAPP algorithm to recognize 101 pattern. A, input plane;
B, output plane.

are loaded into the NS register and, simultaneously, the entire test pattern is shifted another column to the left into the EW register, and C is set to zero.

The NS registers thus contain data that represent the pattern before the second-column shift, and the EW registers contain data that represent the pattern after the second-column shift. The C registers next receive the results from the borrow outputs of the ALUs of each processor element. Also in the fourth step, the EW register is reloaded with the original pattern. The Borrow outputs are equivalent to the logical expression

$$BW = P_{M,N+1} \cdot P_{M,N+2}$$

where $P_{M,N+1}$ is shifted-by-one version of the original pattern in cell row M, column N, and $P_{M,N+2}$ is the shifted-by-two pattern. At the fifth step, the C register gets the Carry output that is logically equivalent to

$$CY = P_{M,N+1} \bullet P_{M,N+1} \bullet P_{M,N+2}$$

In the last step, RAM location 1 gets the final output. It is evident that only those processor elements that are at the left end of a 101 pattern will contain a 1.

Convolution is employed in edge enhancement, for instance, to improve the quality of the image. In convolving a 3 × 3-pixel mask with an 8-bit gray-scale, the mask is placed over every pixel in the image and the product terms in each 3 × 3-pixel window are summed.

Before the GAPP was invented at Martin Marietta-Orlando, image processing techniques requiring the many benefits of systolic array processing were conceptualized but found impractical. Until recently, however, a general purpose, personal computer size, real-time image processing workstation did not exist. The system discussed in this section processes and displays real-time results of preprogrammed GAPP algorithms at RS-170 video rates and is called the GAPP Peripheral Processor (GP²)™.

The architecture used to implement the final GP² system evolved from past Martin Marietta image and signal processor designs with an emphasis on small size, low costs, and display of results in real time. The system is designed to accept Forward Looking Infrared (FLIR), camera or recorded analog data, in a standard RS-170 video format. The system processes this data and provides analog Red/Green/Blue (RGB) or composite signals out to drive a monitor. Processed results are delayed by two frames, but no frames are omitted in processing. Also, no data compression takes place.

The GP² is divided into several functional blocks, as shown in Figure 4.3-4. The system comprises a System Controller (SC) or an interface to a host computer, a Digitizer, a GAPP Controller (GC), an Input Buffer (IB), a GAPP Array,

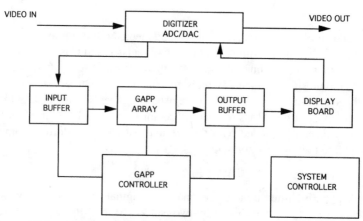

Figure 4.3-4.   GP² Functional block diagram.

an Output Buffer (OB), and a Display Board (DB). These major parts of GP^2 are addressed individually as they occur in the data path. Data formats and communication rates are standard to external devices, but some unique implementations are internally present in the GP^2.

Peak performance figures for the GP^2 with one completely populated array board have been measured as follows.

8 bit addition	4.1 billion additions/s
8 bit multiply, 16 bit result	382.0 million multiplications/s
8 bit 3 × 3 convolution	95.0 million convolutions/s
8 bit 5 × 5 convolution	33.0 million convolutions/s
Binary 21 × 21 correlation	6.7 million correlations/s

The following algorithm performance figures include data I/O instructions and code to process a 192 × 200 image at real-time video rates (Table 4-7). More image data can be processed by adding hardware with no penalty in processing time.

The GP^2 executes at an internal system clock of 10 MHz and results in the execution of a maximum of 330,000 instructions during a video frame time.

TABLE 4-7

ALGORITHM PERFORMANCE FIGURES		
Algorithms	**Number of Instructions**	**Execution Time (10 MHz clock)**
Low pass filter implemented by 3 × 3 convolution	12,922	1.29 ms
Moving target indicator	7118	0.72 ms
Sobel operator	8734	0.87 ms
Template matching 5 × 5 image pixels	6553	0.66 ms

REFERENCES

1. Cloud, Eugene, The Geometric Arithmetic Parallel Processor, Martin-Marietta Corp., Orlando, FL.
2. Davis, Ronald and Thomas, Dave, Systolic Array Chip Matches the Pace of High Speed Processing, *Electronic Design,* Vol. 32, No. 22, October 31, 1984.

3. Smith, W. W. and Sullivan, Paul, Systolic Array Chip Recognizes Visual Patterns Quicker Than a Wink, *Electronic Design*, Vol. 32, No. 24, November 29, 1984.

4. Hord, R. Michael, *Parallel Supercomputing in SIMD Architectures*, CRC Press, Boca Raton, May 1990.

4.4   LORAL ASPRO

Loral's ASPRO-VME is a massively parallel processor. It is ideally suited for applications such as high-speed tactical decision aids, dynamic database management, tracking and correlation, and data fusion applications. ASPRO-VME also has the processing power required for a wide range of image and signal processing applications in support of C^3I (command, control, communications, and intelligence) functions.

ASPRO-VME is the fourth generation architecture derived from Loral Defense Systems-Akron's original massively parallel processor, the STARAN. The second generation ASPRO is deployed on the Navy's E-2C aircraft and the third generation is on CCS-MK2-equipped Los Angeles class fast attack submarines.

ASPRO-VME uses a single-instruction, multiple-data (SIMD) stream architecture and locates data by content, not by address. Content addressability allows for high-speed search operations that are crucial in tracking applications.

The baseline ASPRO-VME configuration occupies only three 6U VME slots, and operates at up to 150 Mflops and 5 Bops of integer arithmetic. ASPRO-VME attains very high performance by the simultaneous processing of single instructions across a large array of processing elements (PEs). The PE array is colocated with its dedicated Module. This array is controlled by a RISC-based Control Processor module working in conjunction with an Array Control module. Each PE is associated with a content addressable multidimensional access (MDA) memory. The content addressability of the memory permits data to be treated as if it were in a large spreadsheet. The MDA memory permits I/O to be performed in the "column" direction with no loss of speed. Users can increase processing power by adding additional Array Memory modules in 6U VME slots. The maximum ASPRO-VME configuration has 16 Array Memory modules that yield 2.4 Gflops and 80 Bops.

The ASPRO-VME is fully programmable in Ada and is backed by a full software development environment.

The open architecture of ASPRO-VME allows easy interface to a variety of host computers such as the HP-750, IBM RS6000, and others.

An associative processor is a computer that contains an associative, or content addressable memory. Associative memories are used to locate entire "words" when subsets of the words are matched against a specific pattern. In ASPRO-VME each "word" contains 128 K bits of data memory. The subset of words that match or relate in a specific way to the set of data patterns are called "responders."

An example of associative processing on a tracking application is illustrated in Figure 4.4-1.

The baseline ASPRO-VME system has three modules: Control Processor (CP), Array Control (AC), and Array Memory (AM). Additionally, ASPRO-VME is expandable up to a total of 16 AM modules.

The CP contains and runs the application programs that control the operation of ASPRO-VME. It does this by sending address and control arguments to the AC and by writing its internal control registers. The AC's internal control registers are directly accessible by the CP as memory mapped registers.

The AC converts the CP's commands into the microcode necessary to perform the operation and transmits the resultant microcode to the attached AMs. System software subroutine libraries are invoked by the Ada application programs to perform the individual array operations.

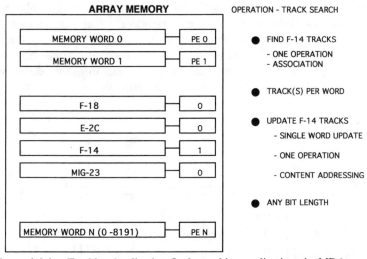

Figure 4.4-1.   Tracking Application. In the tracking application, the MDA memory is laid out as a database that contains a large number of tracks. Each track is in one word of the MDA memory. A search of the database looks for all tracks that contain the attribute "F-14" in a single operation. All tracks that meet the search criteria (association) have their respective PEs set as a responder. In the example, only one F-14 track is found. A subsequent update of the responding track takes place in one operation through content addressing. No a priori knowledge of track location in the MDA memory is required. The attribute length can be any bit length from a single bit to hundreds of bits. A mask inside each PE determines its participation during the operation.

The multidimensional access (MDA) memory located on the AM modules is directly accessible by normal uncached read-write operations performed by the CP or directly from the VMEbus. The MDA memory can be accessed in several different modes.

The AM is memory mapped into the CP's uncached memory address space. This allows the CP to directly access, or perform scalar operations on, data in the AM. Data can also be entered from the VME bus into the AM by direct memory access (DMA) operations.

All sequential processing tasks are performed by the Control Processor. The CP features the R3081E RISC processor with an on-board floating-point co-processor. The CP also handles serial and ethernet communications and networking, and transactions over the VMEbus.

Array operations, or cycles, are performed at a 10-MHz rate. Since the data path within the AM is 1 bit for every PE the internal data rates approach 640 MB/sec within each AM. These high data rates allow floating-point operations to reach a peak of 150 Mflops/AM module. The ASPRO-VME block diagram is shown in Figure 4.4-2.

Application software executes on the ASPRO-VME under the control of Wind River System's VxWorks real-time kernel. VxWorks is widely used, well tested, and comes with very good documentation. Its features include: *multitasking* with preemptive priority scheduling, intertask synchronization and communications facilities, interrupt handling support, watchdog timers, and memory management; and *"Transparent" access* to other VxWorks and UNIX systems via UNIX source-

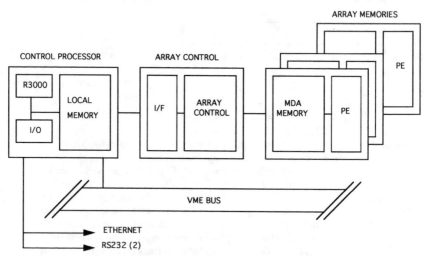

Figure 4.4-2.   The system block diagram shows the CP attached to the VMEbus allowing ASPRO-VME to be accessed by or to access other VME system components.

compatible sockets, remote command execution, remote log in, remote procedure calls (RPC), source-level remote debugging, and remote file access.

The Ada Vectors package was developed by Loral Corporation as an approach to programming parallel algorithms. The vectors package makes use of the ability in Ada to overload operators (i.e., provide unique definitions for addition, subtraction, etc.). The vectors package defines a new data structure called a vector. A vector contains a list of elements. New definitions for all operations involving Vector data types are included. Thus if one wants to add two vector data types together or add a scalar to a vector data type, the vectors package provides the code to the Ada compiler so that the operations can be performed in parallel on the ASPRO-VME.

The Ada vectors package is not a modification of the Ada language. The package supports parallel constructs and is a conventional Ada package much the same as a stack package or a screen management package. The validated VERDIX Ada compiler was not changed.

The Ada vectors package permits the Ada software developer to abstract a problem using vector objects to express parallel operations. The developer has direct control over which operations are parallel and which remain serial; it is not a compiler function. All operations in the Ada Vectors package are logically related to a vector and are independent of serial/parallel CPU considerations.

A Vector is simply a list of numbers or data items (float or integer, character, or Boolean).

Operations do not have to occur for every element in the list. Employing a system-wide (Boolean true or false) mask can indicate which elements participate in the operation and which do not.

Elements of vector lists are only available through vector services. This restriction protects the abstraction of vectors and is consistent with the hardware implementation where parallel vectors are stored in ASPRO-VME's array memory.

Services provided for application developers include creating vectors, loading and unloading vectors, copying vectors, addition, subtraction, multiplication, division, square root, sine, cosine, tangent, filtering primitives, find matches, logical operations, and other arithmetic, logical, and comparison operations between vectors or between vectors and scalars.

The Ada Vectors package uses a collection of library subroutines that implement a set of primitive functions on parallel array memory fields.

The functions provided by the library subroutines include single operand functions (such as simple parallel movement from one field to another, i.e., complement, negate, absolute value, increment, and decrement), integer arithmetic (add, subtract, multiply, divide, reciprocal, and square root), comparisons (exact match, greater than, less than, etc.), bit-wise (or bit-slice-wise) logical operations (and, or, etc.), IEEE floating-point arithmetic and comparisons, scientific functions on IEEE floating-point (sine, natural logarithm, etc.), and some data movement between array memory fields and the Control Processor.

I/O transfers to or from array memory execute at maximum speed when the value in array memory is allocated space aligned on 32-bit address boundaries. This is accomplished using the align command.

The Ada Vectors package contains a library subroutine that examines the ASPRO-VME hardware at run-time to determine the size of array memory. All internal array data structures are set to the array memory size. The number of elements in an array is equal to the number of PEs in the examined hardware (from 512 to 8192 in increments of 512).

The minimum configuration ASPRO-VME is a three-module set that can fit into a standard 6U-220 mm VME enclosure. The set contains one Control Processor (CP) module that provides program execution control over the ASPRO-VME; one Array Control (AC) module that controls the parallel execution of 512 to 8192 processors; and one Array Memory (AM) module that contains 512 PEs. Additional AM modules can be used to linearly increase processing power. Each AM module contains 512 PEs, allowing system configurations of 512 PEs (1 AM module) to 8192 PEs (16 AM modules).

The CP is based on the R3000 ISA architecture. It performs all the scalar processing requirements of the ASPRO-VME. The CP module passes instructions requiring parallel execution (along with data) to the AC.

Array instructions are stored in the CP's memory as library subroutines. The CP fetches these library subroutines and stores the subroutines in the AC. The AC converts the data to parallel instructions that are broadcast parallel instructions to each AM module for simultaneous execution by the PEs on data attributes in their local memory.

The heart of ASPRO-VME's processing power is the AM module. Each AM module has 512 PEs with performance of 150 Mflops and 5 Bops integer operations. Each PE has a local memory of 128K bits. Bandwidth between the PEs and memory is 640 MB/sec for each AM. Additional AM modules can be added to linearly increase processing power.

One key feature of ASPRO-VME is its ability to directly access the MDA memory from an external source on the VMEbus at VMEbus speeds. Another version of the AM module, called the High Speed Input/Output Array Memory (HSIO AM), has a high-speed I/O interface can provide a total of 90 MB/sec transfer rate into and out of each AM module for direct sensor data input, or display data output.

The minimum three-module set provides 150 Mflops of 32-bit single-precision, IEEE-format, floating-point performance, and 5 Bops of 32-bit integer add performance. The basic cycle time of the ASPRO-VME is 100 nanoseconds. Expansion and increased performance are provided by adding AM modules. The maximum 18-module set contains 16 AMs, which yields 2.4 Gflops and 80 Bops, and fits in a standard 19-inch rack-mounted unit, 12.24 inches high and 22.5 inches deep.

In ASPRO-VME, processing elements (PEs) are arranged in a two-dimensional layout, with each PE having 128 K bits of associated memory on each AM module. A fully populated ASPRO-VME would contain 8192 PEs. The memory address allocation for PE supports an expansion of up to 1 million bits/PE. Multi-dimensional access permits access in either row or column mode.

All PEs that respond to associative searches will set a bit within the PE. Each AM forms the logical OR of the bits that are set on the module, which is sent back to the AC as the module SUMOR.

The final resolver obtains SUMOR information from each of the AMs to determine the first responder's address. It then provides the most significant part of the resolved address to enable the first responding AM. The AM, thus enabled, utilizes its on-board resolver to complete the resolved address for content addressing operations.

The PEs form the heart of ASPRO-VME. Each PE contains an I/O register, mask register, response store registers, a 32-bit parallel arithmetic and logic unit (PALU), and eight file registers to store intermediate arithmetic results. They provide all the processing power, store the results of associative searches, and provide the content addressability of ASPRO-VME. The PEs are grouped into blocks of 32. Each block uses one flip network. One single AM module contains 16 such groupings for a total of 512 PEs. The group of 32 PEs and the flip network are contained in two identical ASICs. Each par of ASICs provides a 32-bit interface to its 32 MDA memory words, and a 32-bit interface to the direct I/O bus from AC.

Chapter 5

MIMD

Nine commercial multiple-instruction-multiple-datastream parallel supercomputers are examined in detail in this chapter: Paragon, Connection Machine-5, IBM SP1 and SP2, Kendall Square KSR1, Cray T3D, Parsytec, Convex, nCUBE, and SGI Power Challengearray.

These have been selected from among many as the most illustrative of the forces that have driven the evolution of MIMD supercomputers. The order of presentation was chosen to facilitate the exposition.

5.1 THE INTEL PARAGON™ XP/S SYSTEM

The Paragon system (Figure 5.1-1) employs a scalable, heterogeneous multicomputer architecture [1, 2]. Its various processing nodes populate a two-dimensional interconnection network. Nodes provide a variety of services including computation, input/output services, operating system services, and network connectivity (see Figure 5.1-2). Main memory is physically distributed among the nodes.

The system makes extensive use of microprocessor technology. Each employs at least two Intel i860 XP processors: one or more application processors dedicated to executing a user's application program, and a message processor dedicated to the task of internode communication, the feature responsible for providing low latency and high bandwidth between every node in the system. Two nodes, anywhere in a Paragon system, achieve process-to-process transfer latency of 25 microseconds—regardless of the size and configuration of the system. Messages are routed automatically, so the application programmer sees a system in which every node is connected to every node, with no difference in performance among nodes.

The key to high sustained performance in a multicomputer is balanced design: The speed and memory capacity of the individual nodes must be matched by

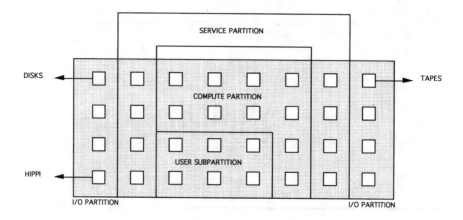

Figure 5.1-1.   The architecture of the Paragon system. All programming models
are supported because the topology of the interconnect network
allows the programmer to ignore the physical location and function
of the nodes. Systems are divided into partitions, which can consist
of as little as one node or as much as all of the nodes. In this example, the system is configured with eight nodes dedicated to I/O
interfaces, eight nodes to system services, and the remainder to the
user applications.

the system's interconnection network, mass storage facilities, graphics devices,
and network connections. The design of the Paragon XP/S system is balanced in
all these areas.

Within each node, the microprocessor speed is matched by on-chip cache
performance and high-bandwidth memory units. Aggregate interconnect bandwidth scales with the number of nodes and supports the efficient operation of
systems with many thousands of processors. The bandwidth and latency of the
node-to-network interfaces and node-to-node communication channels are carefully matched to the node memory bandwidth and execution speeds. Aggregate
I/O performance, like computational performance, scales with the number of I/O
nodes. The I/O interfaces are custom engineered to the needs of specific industry-
standard bus and channel interfaces. With greater bandwidth than virtually any
peripheral device, point-to-point I/O channel, or local area network, the interconnect sustains this I/O performance from or to any node in the system.

Balance within each Paragon node is the result of engineering the memory
unit, messaging facilities, and external I/O interfaces to match the performance
demands of the Intel i860 XP microprocessor. Capable of 42 MIPS and 75
double-precision MFLOPS when operating at 50 MHz (20 ns clock cycle), the
2.55-million-transistor Intel i860 XP provides four-way set associative, on-chip,

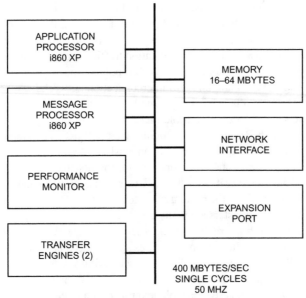

Figure 5.1-2. The Paragon node is modular in architecture and implementation, with the network interconnect boosted with a dedicated Intel i860 XP separate from the i860 XP processor used for user applications. The design also benefits from a performance monitor with ties to all the major components and a connection to its own (optional) data collection network.

16-Kbyte instruction and data caches. Floating-point-unit-to-cache bandwidth peaks at 800 Mbytes/second. Cache refills to local memory take place at 400 Mbytes/sec and are fully supported by an interleaved, dual-bank, single-cycle, 64-bit memory unit design. Local memory is based on 60 ns DRAM arrays with full single-bit error correction and double-bit error detection.

Interconnect performance is fundamentally determined by the number of wires that cross from one half of the network (the bisection) to the other. Networks with more wires, assuming they are equally short, should be faster than networks with fewer wires. In practice, the number of wires crossing the bisection is limited by electrical factors, such as power dissipation, and mechanical factors, such as trace, connector, and cable density. Designing an optimal network is principally an effort to make the most efficient use of the available wires, given that the wire limit is essentially the same at a given moment in time for all practical designs.

Extensive analytical studies, thousands of hours of simulation, and several prototype systems (most notably the Touchstone DELTA and SIGMA systems) have demonstrated that low-dimensional mesh networks make the most efficient use of available wires. That means, given an equal number of wires at the bisection of the network, a two-dimensional mesh (Figure 5.1-3) will outperform any toroidal, hypercube, or tree-structured network for uniformly distributed communication traffic in systems containing up to several thousand nodes. The mesh advantage actually grows as the communication traffic becomes more localized.

To deliver on the promise of the mesh topology, the switching elements and electrical design of the Paragon XP/S interconnection network have been carefully considered. The individual switches, located at each vertex in the network, are fabricated in the same 0.75 micron, triple-metal, CMOS technology used to build the Intel i860 XP. Each Paragon Mesh Routing Chip (PMRC, also sometimes called iMRC due to its i860 origins) can simultaneously route traffic moving in the four network directions ($\pm$ X and $\pm$ Y), and to and from its attached node, at speeds in excess of 200 Mbytes/sec. It takes the PMRC less than 40 ns to make an individual routing decision and close the necessary switches. Internal operation of the router is fully parallel, and all transfers are parity checked. To maximize router bandwidth, it is fully pipelined both internally and externally. Message traffic moving from one router to another is pipelined along the wires so that speed becomes independent of distance for all practical purposes. The routing algorithm is provably deadlock-free.

Instead of a passive backplane, as used in most conventional systems, the Paragon system uses an active backplane built from a rectangular arrangement of

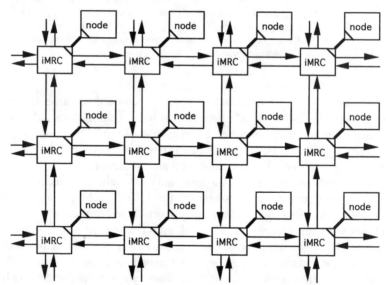

Figure 5.1-3.   The Paragon architecture utilizes a two-dimensional mesh network topology, originally developed under the Intel-DARPA Touchstone project. Every node, regardless of its function in the system, utilizes the interconnection network, permitting every system service and function to scale with the system as it grows to more than a 1000 nodes.

router chips. Each backplane section connects to its horizontal and vertical neighbors by flexible printed circuits and ultrareliable connectors. Open the back door of a Paragon system and the backplanes are completely clean. Gone are the fragile cables and wiring mat that have become the trademarks of conventional vector supercomputers and are still found in many highly parallel machines. Also missing are the critically measured and cut clock control lines, as the Paragon interconnect is completely self-timed and clock-free. Only the Paragon mesh routers are visible as they go about the business of moving information between the nodes at speeds limited only by their underlying VLSI technology and the speed of light. For systems with as few as 256 nodes, the bisection bandwidth is in excess of 6 Gbytes per second. A Paragon system with 1000 nodes has a bisection bandwidth of over 12 Gbytes per second.

A semicustom VLSI chip, the Network Interface Controller (NIC), provides a full-bandwidth, pipelined electrical interface between a node and its PMRC. Included in this chip is a parity-checked, full-duplex router port and end-to-end error checking for each message transfer. Internal parallel operation permits simultaneous inbound and outbound communication at 200 Mbytes/sec. Since the node processor cannot sustain this speed for long messages, two block transfer engines are designed into the memory interface. They can move up to 4096 bytes of data at a time at an aggregate speed just short of the full node memory bandwidth.

To complete the node-network interface, a second full-speed, full function Intel i860 XP message processor is also incorporated into the basic node design. Its purpose is to free the application processor from the details of internode communication and to provide a variety of communication services within the Paragon system. Operating in parallel, the message processor handles all communication traffic to and from the node. It autonomously receives messages and prepares them for use by the application processor. Similarly, it handles all the details of sending a message, including protocol and packetization, if required.

The message processor has many subtle benefits that become obvious only after understanding the low-level issues of MIMD multicomputer design. By freeing the application processor of messaging details, the message processor reduces cache turbulence in both processors. The application code and data stay in the caches on board the application processor to sustain performance. The messaging software of the operating system is small enough to remain resident in the message processor's code and data caches. This guarantees the lowest possible latency, because library code does not first have to fill the processor caches before it can run at full speed. The dual-processor design also avoids a costly context switch in the application processor during messaging operations and while handling interrupts from the network interface. This, in turn, helps avoid draining the floating-point and memory pipelines in the application processor.

The message processor at each node also enables the Paragon system to offer a full range of global operations that are performance-critical in both MIMD and

SPMD applications. The global operations that are autonomously implemented by cooperating message processors include broadcasting to and synchronizing a group of compute nodes, as well as mathematical operations, such as global sum and global minimum, for both integer and floating-point operands. Global operations are also available for logical operands, such as global AND and OR, and for aggregates, such as global string concatenation.

Paragon XP/S configurations offer peak computational speeds ranging from 5 to 300 GFLOPS. This computational performance is matched by a high-bandwidth low-latency interconnection network that passes messages between any two nodes in the system at rates of 200 Mbytes/sec (full duplex).

Every aspect of the Paragon architecture is scalable. Main storage capacity scales to 128 Gbytes of dynamic RAM and more than a terabyte of high speed internal disk storage. Peripherals and network interfaces are scalable—multiple channels for SCSI-2, VMI, Ethernet, and High Performance Parallel Interface (HIPPI) to meet any I/O requirement. The bandwidth of the system's interconnection network rises with the number of nodes in the system.

The Paragon operating system, a full implementation of UNIX, also scales. Once applications are running on the Paragon system, they too become scalable. For example, a computational fluid dynamics code could be developed and tested on a small number of nodes. When ready for production, the number of nodes used for the production runs can be based on the desired performance—400 nodes or 1000 nodes could be employed without change of the application.

The system can scale to match the size of the data set without changing the application or the time it takes to get the answer. For instance, often the same algorithm can be used for a larger data set, allowing the user to model an entire structure rather than just a portion of the structure. Using the same application on a larger number of nodes allows the use of these larger data sets while retaining the quick turnaround time of the answer.

Paragon's transparently distributed UNIX operating system—the first full UNIX implementation for a massively parallel supercomputer—ensures flexibility in managing the Paragon system, developing applications, and integrating the system into heterogeneous distributed computing environments. The operating system is the result of developments in distributed operating system services and microkernel technology, and delivers full UNIX services to every node in the machine.

Adding to the system's ease of use and further integrating parallel supercomputing into the technical computing environment, the Paragon system supports client/server computing via scalable HIPPI, FDDI, and Ethernet networking. For flexible system access, users can submit jobs over a network or can log directly onto any Paragon node. The Paragon system can be used for batch and interactive jobs simultaneously, and can be reallocated dynamically to adjust the proportion of interactive and batch services. Comprehensive resource control utilities are available for allocating, tracking, and controlling system resources.

TABLE 5-1

THE PARAGON™ XP/S SYSTEM AT A GLANCE	

Capacity	5–300 GFLOPS peak 64-bit floating-point performance
	2.8–160 KMIPS peak integer performance
	Node-to-node message routing at 200 Mbyte/sec (full duplex)
	1–128 Gbytes main memory, up to 500 Gbyte/sec aggregate
	   bandwidth
	2–128 Mbytes processor cache, up to 4.0 Tbytes/sec aggregate
	   bandwidth
	6 Gbytes to 1 Tbyte internal disk storage, up to 6.4 Gbyte/sec
	   aggregate I/O bandwidth

System Architecture	Scalable, distributed-memory multicomputer MIMD control model

Node
   Architecture
	Nodes based on Intel's 50 MHz i860™ XP processor
	75 double-precision MFLOPS, 42 VAX MIPS peak performance
	   per processor
	16–128 Mbytes DRAM per node
	400 Mbytes/sec processor-to-memory bandwidth
	800 Mbyte/sec processor-to-cache bandwidth

Interconnect Architecture	2D mesh topology Pipelined, hardware-based internode communications
Operating System	Full UNIX operating system Transparent, distributed, scalable services Conformance with POSIX, System V.3, 4.3bsd Virtual memory
System Access	Simultaneous batch and interactive operation NQS, MACS utilities for resource management Client/server access, direct logins, remote host
Connectivity	Multiple HIPPI channels with 100 Mbytes/sec bandwidth each Multiple Ethernet channels Multiple VME* connections NFS, TCP/IP, DECnet* protocols, UniTree client support
Programming Environment	C, FORTRAN, Ada, C++, Data-parallel FORTRAN Integrated tool suite with a Motif-based GUI FORGE and CAST parallelization tools Intel ProSolver parallel equation solvers BLAS, NAG, SEGlib, and other math libraries Interactive Parallel Debugger (IPD) Hardware-aided Performance Visualization System (PVS) Operating system support for shared virtual memory
Visualization Tools	X Window System PEX Distributed Graphics Library (DGL) client support AVS and Explorer interactive visualizers Connectivity to HIPPI frame buffers

Intel's MIMD architecture supports all programming styles and paradigms. Applications can be developed using the programmer's preferred programming model—object-oriented, Single Program Multiple Data (SPMD), Single Instruction Multiple Data (SIMD), Multiple Instruction Multiple Data (MIMD), Shared Memory, or Vector Shared Memory.

Paragon parallel computer-aided software engineering (CASE) tools assist application developers in writing and porting applications to fully realize the system's performance potential. The development tool suite is integrated through a common database and a Motif-based graphical user interface. Tools include optimizing compilers for FORTRAN, C, C++, Ada, and Data Parallel FORTRAN; the Interactive Parallel Debugger (IPD); parallelization tools such as FORGE and CAST; Intel's highly tuned ProSolver library of equation solvers; Intel's Performance Visualization System (PVS™); and the Performance Analysis Tools (iPAT™).

The Paragon system's massively parallel architecture provides performance and flexibility beyond the reach of traditional vector supercomputers (Table 5-1).

- Transparent software scalability. Once an application is running on the Paragon XP/S system, it can run without change on any number of nodes. Users can move back and forth from a 5 GFLOP to a 300 GFLOP performance level with no reprogramming or system reconfiguration.

- Simultaneous interactive and batch operations. The Paragon system does not force the users to decide between operating in batch mode or interactive mode—all modes of operation are provided simultaneously and all system resources can be dynamically reallocated as the system is running.

- Native or networked development. Developers are given the option to choose software tools directly on the system in native mode or on their own workstations, downloading their applications.

REFERENCES

1. Paragon XP/S Product Overview, Intel Corp., Supercomputing Systems Systems Division, 15201 N.W. Greenbrier Parkway, Beaverton, OR, 1991.

2. Hord, R. Michael, *Parallel Supercomputing in MIMD Architectures,* CRC Press, Boca Raton, March 1993.

5.2   CONNECTION MACHINE-5

A CM-5 system (Figure 5.2-1) may contain hundreds or thousands of parallel processing nodes [1]. Each node has its own memory. Nodes can fetch from the same address in their respective memories to execute the same (SIMD-style) instruction, or from individually chosen addresses to execute independent (MIMD-style) instructions.

The processing nodes are supervised by a control processor, which runs an enhanced version of the UNIX operating system. Program loading begins on the control processor; it broadcasts blocks of instructions to the parallel processing nodes and then initiates execution. When all nodes are operating on a single control thread, the processing nodes are kept closely synchronized and blocks are broadcast as needed. (There is no need to store an entire copy of the program at each node.) When the nodes take different branches, they fetch instructions independently and synchronize only as required by the algorithm under program control.

To maximize system usefulness, a system administrator may divide the parallel processing nodes into groups, known as partitions. There is a separate control processor, known as a partition manager, for each partition. Each user process executes on a single partition, but may exchange data with processes on other partitions. Since all partitions utilize UNIX timesharing and security features, each allows multiple users to access the partition while ensuring that no user's program interferes with another's.

Other control processors in the CM-5 system manage the system's I/O devices and interfaces. This organization allows a process on any partition to access any I/O device, and ensures that access to one device does not impede access to other devices.

Every control processor and parallel processing node in the CM-5 is connected to two scalable interprocessor communication networks, designed to give low latency combined with high bandwidth in any possible configuration a user may wish to apply to a problem. Any node may present information, tagged with its logical destination, for delivery via an optimal route. The network design provides low latency for transmissions to near neighboring addresses, while preserving a high, predictable bandwidth for more distant communications.

The two interprocessor communications networks are the Data Network and the Control Network. In general, the Control Network is used for operations that involve all the nodes at once, such as synchronization operations and broadcasting; the Data Network is used for bulk data transfers where each item has a single source and destination.

A third network, the Diagnostics Network, is visible only to the system administrator; it keeps tabs on the physical well-being of the system.

External networks, such as Ethernet and FDDI, may also be connected to a CM-5 system via the control processors.

The CM-5 runs a UNIX-based operating system; it provides its own high-speed parallel file system, and also allows full access to ordinary NFS file systems. It supports both HIPPI (high-performance parallel interface) and VME interfaces, thus allowing connections to a wide range of computers and I/O devices, while using standard UNIX commands and programming techniques throughout.

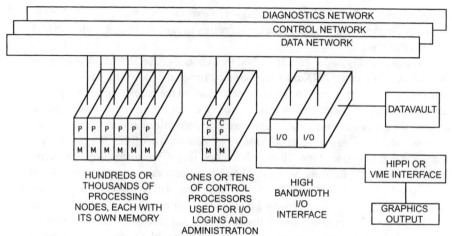

Figure 5.2-1.   Organization of the Connection Machine system. Functionally, the
CM-5 is divided into three major areas. The first contains some
number of partitions, which manage and execute user applications;
the second contains some number of I/O devices and interfaces;
and the third contains the two interprocessor communication
networks that connect all parts of the first two areas. Because all
areas of the system are connected by the Data Network and the
Control Network, all can exchange information efficiently. The
two networks provide high bandwidth transfer of messages of all
sorts: downloading code from a control processor to its nodes,
passing I/O requests and acknowledgments between control
processors, and transferring data, either among nodes (whether in a
single partition or in different partitions) or between nodes and I/O
devices.

A CMIO interface supports mass storage devices such as the DataVault and enables sharing of data with CM-2 systems.

I/O capacity may be scaled independently of the number of computational
processors. A CM-5 system of any size can have the I/O capacity it needs,
whether that be measured in local storage, in bandwidth, or in access to a variety
of remote data sources. Communications capacity scales both with processors
and with I/O.

Just as every partition is managed by a control processor, every I/O device is
managed by an input/output control processor (IOCP), which provides the software that supports the file system, device driver, and communications protocols.
Like partitions, I/O devices and interfaces use the Data Network and the Control
Network to communicate with processes running in other parts of the machine. If
greater bandwidth is desired, files can be spread across multiple I/O devices: A
striped set of eight DataVaults, for example, can provide eight times the I/O bandwidth of a single DataVault.

The same hardware and software mechanisms that transfer data between a partition and an I/O device can also transfer data from one partition to another (through a named UNIX pipe) or from one I/O device to another.

The architecture of the CM-5 is optimized for data parallel processing of large, complex problems. The Data Network and Control Network support fully general patterns of point-to-point and multi-way communication, yet reward patterns that exhibit good locality (such as nearest-neighbor communications) with reduced latency and increased throughput. Specific hardware and software support improve the speed of many common special cases.

A Connection Machine Model CM-5 system contains thousands of computational processing nodes, one or more control processors, and I/O units that support mass storage, graphic display devices, and VME and HIPPI peripherals. These are connected by the Control Network and the Data Network. (For a high-level sketch of these components, see Figure 5.2-2.)

Every processing node is a general-purpose computer that can fetch and interpret its own instruction stream, execute arithmetic and logic instructions, calculate

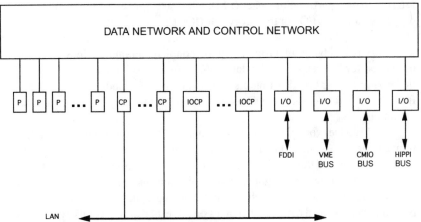

Figure 5.2-2.   System components. A CM-5 system contains tens, hundreds, or thousands of processing nodes, each with up to 128 Mflops of 64-bit floating-point performance. It also contains a number of I/O devices and external connections. The number of I/O devices and external connections is independent of the number of processing nodes. Both processing and I/O resources are managed by a relatively small set of control processors. All these components are uniformly integrated into the system by two internal communications networks, the Control Network and the Data Network. The Control Network provides multiway operations that coordinate thousands of participants, while the Data Network supports high-bandwidth bulk data transfers. The capacity of each network scales up with the size of the system; every processing node or I/O device gets the network capacity it needs.

memory addresses, and perform interprocessor communication. The processing nodes in a CM-5 system can perform independent tasks or collaborate on a single problem. Each processing node has 8, 16, or 32 Mbytes of memory; with the high-performance arithmetic accelerator, it has the full 32 Mbytes of memory and delivers up to 128 Mips or 128 Mflops.

The control processors are responsible for administrative actions such as scheduling user tasks, allocating resources, servicing I/O requests, accounting, enforcing security, and diagnosing component failures. In addition, they may also execute some of the code for a user program. Control processors have the same general capabilities as processing nodes but are specialized for performing managerial functions rather than computational functions. For example, control processors have additional I/O connections and lack the high-performance arithmetic accelerator (Figure 5.2-3).

In a small system, one control processor may play a number of roles. In larger systems, individual control processors are often dedicated to particular tasks and referred to by names that reflect those tasks. Thus, a control processor that manages a partition and initiates execution of applications on that partition is referred to as a partition manager (PM), whereas a processor that controls an I/O device is called an I/O control processor (IOCP).

The Control Network provides tightly coupled communications services. It is optimized for fast response (low latency). Its functions include synchronizing the processing nodes, broadcasting a single value to every node, combining a value from every node to produce a single result, and computing certain parallel prefix operations.

The Data Network provides loosely coupled communications services. It is optimized for high bandwidth. Its basic function is to provide point-to-point data delivery for tens of thousands of items simultaneously. Special cases of this functionality include nearest-neighbor communication and FFT butterflies. Communications requests and data delivery need not be synchronized. Once the Data Network has accepted a message, it takes on all responsibility for its eventual delivery; the sending processor can then perform other computations while the message is in transit. Recipients may poll for messages or be notified by interrupt on arrival. The Data Network also transmits data between the processing nodes and I/O units.

A standard Network Interface (NI) connects each node or control processor to the Control Network and Data Network. This is a memory-mapped control unit; reading or writing particular memory addresses will access network control registers or trigger communication operations.

The I/O units are connected to the Control Network and Data Network in exactly the same way the processors, using the same Network Interface. Many I/O devices require more data bandwidth than a single NI can provide; in such cases multiple NI units are ganged. For example, a HIPPI channel interface contains 6 NI units, which provide access to 6 Data Network ports. (At 20 Mbytes/sec

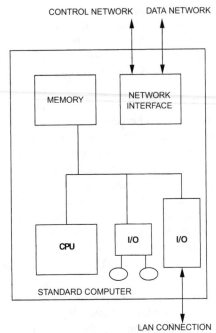

Figure 5.2-3.   Control processor. The basic CM-5 control processor consists of
a RISC microprocessor, memory subsystem, I/O (including local
disks and Ethernet connections), and a CM-5 Network Interface, all
connected to a standard 64-bit bus. Except for the Network Interface,
this is a standard off-the-shelf workstation-class computer system.
The Network Interface connects the control processor to the rest of
the system through the Control Network and Data Network. Each
control processor runs CMost, a UNIX-based operating system with
extensions for managing the parallel-processing resources of the
CM-5. Some control processors are used to manage computational
resources and some are used to manage I/O resources.

apiece, 6 NI units provide enough bandwidth for a 100 Mbyte/sec HIPPI interface
with some to spare.)

Individual I/O devices are controlled by dedicated I/O control processors
(IOCP). Some I/O devices are interfaces to external buses or networks; these include interfaces to VME buses and HIPPI channels. Noteworthy features of the
I/O architecture are that I/O and computation can proceed independently and in
parallel, that data may be transferred between I/O devices without involving the
processing nodes, and that the number of I/O devices may be increased completely independently of the number of processing nodes.

Lurking in the background is a third network, the Diagnostic Network. It can
be used to isolate any hardware component and to test both the component itself

and all connections to other components. The Diagnostic Network pervades the hardware system but is completely invisible to the user; indeed, it is invisible to most of the control processors. A small number of the control processors include command interfaces for the Diagnostic Network; at any given time, one of these control processors provides the System Console function.

The virtual machine provided by the hardware and operating system to a single user task consists of a control processor acting as a partition manager (PM), a set of processing nodes, and facilities for interprocessor communication. Each node is an ordinary general-purpose microprocessor capable of executing code written in C, Fortran, or assembly language. The processing nodes may also have optional vector units for high arithmetic performance.

The operating system is CMost, a version of SunOS enhanced to manage CM-5 processor, I/O, and network resources. The PM provides full UNIX services through standard UNIX system calls. Each processing node provides a limited set of UNIX services.

A user task consists of a standard UNIX process running on the PM and a process running on each of the processing nodes. Under timesharing, all processors are scheduled *en masse*, so that all are processing the same user task at the same time. Each process of the user task, whether on the PM or on a processing node, may execute completely independently of the rest during their common time slice.

The Control Network and Data Network allow the various processes to synchronize and transfer data among themselves. The unprivileged control registers of the network interface hardware are mapped into the memory space of each user process, so that user programs on the various processors may communicate without incurring any operating system overhead.

Each process of a user task can read and write messages directly to the Control Network and the Data Network. The network used depends on the task to be performed.

The Control Network (CN) is responsible for communications patterns in which many processors may be involved in the processing of each datum. One example is broadcasting, where one processor provides a value and all other processors receive a copy. Another is reduction, where every processor provides a value and all values are combined to produce a single result. Values may be combined by summing them, finding the maximum input value, or taking the logical OR or exclusive OR of all input values; the combined result may be delivered to a single processor or to all processors. (Software provides minimum-value and logical AND operations by inverting the inputs, applying the hardware maximum-value or logical OR operation, then inverting the result.) Note that the control processor does not play a privileged role in these operations; a value may be broadcast from, or received by, the control processor or any processing node with equal facility.

The Control Network contains integer and logical arithmetic hardware for carrying out reduction operations. This hardware is distinct from the arithmetic hardware of processing nodes; CN operations may be overlapped with arithmetic processing by the processors themselves. The arithmetic hardware of the Control Network can also compute various forms of parallel prefix operations, where every processor provides a value and receives a result; the nth result is produced by combining the first n input values. Segmented parallel prefix operations are also supported in hardware.

The Control Network provides a form of two-phase barrier synchronization (also known as "fuzzy" or "soft" barriers). A processor can indicate to the Control Network that it is ready to enter the barrier. When all processors have checked in, the Control Network relays this fact to all processors. A processor can thus overlap unrelated processing with the possible waiting period between the time it has checked in and the time it has been determined that all processors have checked in. This allows thousands of processors to guarantee the ordering of certain of their operations without ever requiring that they all be exactly synchronized at one given instant.

The Data Network is responsible for reliable, deadlock-free point-to-point transmission of tens of thousands of messages at once. Neither the senders nor the receivers of messages need be globally synchronized. At any time, any processor may send a message to any processor in the user task. This is done by writing first the destination processor number, and then the data to be sent, to control registers in the Network Interface (NI). Once the Data Network has accepted the message, it assumes all responsibility for eventual delivery of the message to its destination. In order for a message to be delivered, the processor to which it was sent must accept the message from the Data Network. However, processor resources are not required for forwarding messages. The operation of the Data Network is independent of the processing nodes, which may carry out unrelated computations while messages are in transit.

There is no separate interface for special patterns of point-to-point communication, such as nearest neighbors within a grid. The Data Network presents a uniform interface to the software. The hardware implementation, however, has been tuned to exploit the locality found in commonly used communication patterns.

Data Network performance follows a simple model. The Data Network provides enough bandwidth for every Network Interface to sustain data transfers at 20 Mbytes/sec to any other NI within its group of 4; at 10 Mbytes/sec to any other NI within its group of 16; or at 5 Mbytes/sec to any other NI in the system. (Two Network Interfaces are in the same group of $2k$ if their network addresses differ on the k lowest-order bits.) These figures are for maximum sustained network hardware performance, which is sufficient to handle the transfer rates sustainable by node software. Note that worst-case performance is only a factor of 4 worse than best-case performance. Other network designs have much larger worst/best ratios.

To see the consequences of this performance model, consider communication within a two-dimensional grid. If, say, the processors are organized so that each group of 4 represents a 2 × 2 patch of the grid, and each group of 16 processors represents a 4 × 4 patch of the grid, then nearest-neighbor communication can be sustained at the maximum rate of 20 Mbytes/sec per processor. For within each group of four, two of the processors have neighbors in a given direction (North, East, West, South) that lie within the same group, and therefore can transmit at the maximum rate. The other two processors have neighbors outside the group of four. But the Data Network provides bandwidth of 40 Mbytes/sec out of that group, enough for each of the four processors to achieve 10 Mbytes/sec within a group of 16. The same argument applies to the four processors in a group of 16 that have neighbors outside the group: Not all processors have neighbors outside the group, so their outside-the-group bandwidth can be borrowed to provide maximum bandwidth to processors that do have neighbors outside the group.

There are two mechanisms for notifying a receiver that a message is available. The arrival of a message sets a status flag in a Network Interface control register; a user program can poll this flat to determine whether an incoming message is available. The arrival of a message can also optionally signal an interrupt. Interrupt handling is a privileged operation, but the operating system converts an arrived-message interrupt into a signal to the user process. Every message bears a four-bit tag; under operating system control, some tags cause message-arrival interrupts and others do not. (The operating system reserves certain of the tag numbers for its own use; the hardware signals an invalid-operation interrupt to the operating system if a user program attempts to use a reserved message tag.)

The Control Network and Data Network provide flow control autonomously. In addition, two mechanisms exist for notifying a sender that the network is temporarily clogged. Failure of the network to accept a message sets a status flag in a Network Interface control register; a user program can poll this flag to determine whether a retry is required. Failure to accept a message can also optionally signal an interrupt.

Data can also be transferred from one user task to another, or to and from I/O devices. Both kinds of transfer are managed by the operating system using a common mechanism. An intertask data transfer is simply an I/O transfer through a named UNIX pipe.

REFERENCE

1. The Connection Machine CM-5 Technical Summary, Thinking Machines Corp., Cambridge, MA, Oct. 1991.

5.3   IBM SP1 AND SP2

IBM's initial offering, the Scalable POWERparallel 1 (SP1), established IBM as a competitor in high-performance computing systems (Figure 5.3-1). Building on

SP1 System Overview

	PER NODE	PER FRAME	PER SYSTEM
PROCESSORS	1	16	64
PEAK PERFORMANCE	125 MF	2 GF	8 GF
MEMORY	64-256 MB	1-4 GB	4-16 GB
STORAGE	0-2 GB	0-32 GB	0-128 GB

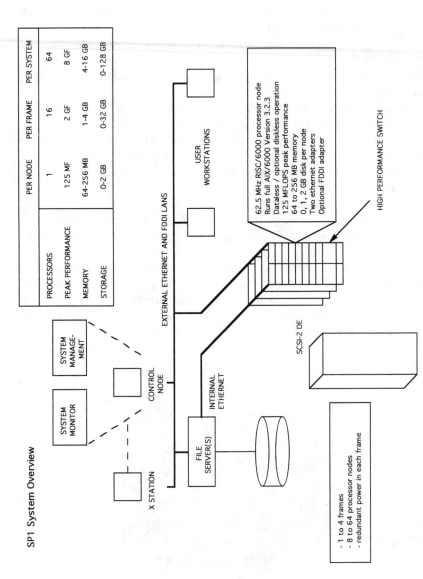

SYSTEM MONITOR

SYSTEM MANAGE-MENT

X STATION

CONTROL NODE

FILE SERVER(S)

INTERNAL ETHERNET

EXTERNAL ETHERNET AND FDDI LANS

USER WORKSTATIONS

62.5 MHz RISC/6000 processor node
Runs full AIX/6000 Version 3.2.3
Dataless / optional diskless operation
125 MFLOPS peak performance
64 to 256 MB memory
0, 1, 2 GB disk per node
Two ethernet adapters
Optional FDDI adapter

HIGH PERFORMANCE SWITCH

SCSI-2 DE

- 1 to 4 frames
- 8 to 64 processor nodes
- redundant power in each frame

Figure 5.3-1. SP1 system overview.

103

the experience of the SP1, the IBM Scalable POWERparallel 2 (SP2) is a more powerful and cost-effective solution for technical enterprises.

As IBM's second offering in the POWERparallel family, the SP2 system is designed to handle the very compute and I/O intensive jobs, allowing one to address projects and problems once considered too complex or expensive to pursue. Although the SP2 system maintains the SP1's architecture and compute-intensive application support, it now provides a UNIX-based platform for database query, online transaction processing, business management, and batch business applications as well.

The SP2, based on the POWER2 chip, has a scalable architecture that allows expansion over time to meet changing computing requirements. Systems range in size from four processor nodes, to a maximum standard configuration of 128 and beyond on special request. The SP2 integrates the POWER2 microprocessor, IBM's RISC system/6000 technology, which approximately doubles the performance of the SP1.

Since the SP2 is based on the AIX/6000 operating system and RISC System/6000 POWER2 technology, thousands of RISC System/6000 applications will run, unchanged, on the SP2. This means that applications integrate to the SP2 easily and quickly, without the need for reprogramming.

The SP2 software lets you run various combinations of serial, parallel, interactive, and batch jobs concurrently (Figure 5.3-2). This allows a site to fully utilize the computing power of the SP2 as processing needs change and grow. In addition, a site can easily change the partitioning scheme of the system to suit new processing requirements.

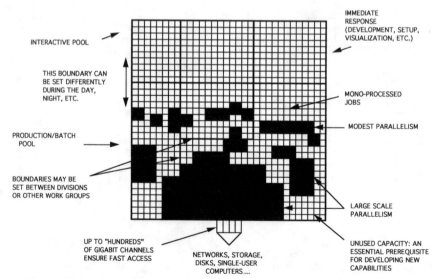

Figure 5.3-2.   Flexible architecture.

Although the SP2 can handle complex jobs, simplicity is emphasized. The AIX Parallel Systems Support Programs (PSSP) software, provided with the SP2, allows one to operate and control the SP2 from a single workstation console.

Two different types of SP2 processor nodes are available; thin nodes and wide nodes. Thin nodes are best suited as the compute nodes. Thin nodes offer the flexibility of running any combination of interactive, batch, serial, and parallel jobs simultaneously, so the SP2 can be used for a diverse set of computing scenarios. Thin nodes may be shared by any one of these jobs or may be dedicated to running a single application.

Wide nodes are generally configured to act as servers; they can provide the various services that are required to run the jobs. Wide nodes integrate expanded memory and directly attached storage device options into an SP2 system. They also support a wide variety of microchannel adapters, allowing the SP2 to handle a large number of compute-intensive applications.

An SP2 system frame houses up to 16 thin processor nodes or up to eight wide nodes. Frames connect together for a maximum standard configuration of 128 processor nodes. Configurations also mix and match thin and wide nodes together in a single frame.

The SP2 frame contains redundant power supplies so that if one fails, another takes over. The frame is also designed for concurrent maintenance; each processor node can be removed and repaired without interrupting operations on the other nodes. As a result, the SP2 provides a highly reliable system. In addition, the SP2 comes with a built-in data modem for automated "phone home" maintenance.

A High-Performance Switch is also available with the SP2 that allows low latency, high-bandwidth communication within the system for optimum parallel execution. It provides the interconnection network that allows all processors in the system to send messages to one another simultaneously. The SP2 switch is a multistage network; a site can add switches to scale the system upward while continuing to provide the same level of bandwidth to each processor node. The switch network can speed up TCP/IP, file transfers, remove procedure calls, and Relational Database functions.

A single switch supports up to 16 processor nodes, so each SP2 frame accommodates one switch. The Scalable POWERparallel Systems family continues the AIX and RISC System/6000 policy of adherence to open systems standards including connection to I/O devices, networks of workstations, and mainframe networks. Ethernet, HIPPI, SCSI, FDDI, and Token Ring interfaces are supported by the SP2.

A site can also connect the SP2 to host systems by means of the BMCA adapter and the ESCON channel interface. This allows using S/390 files and devices from the SP2.

IBM's software offerings for the SP2 are an interlocking set of program products. They work together to address a wide range of system and application needs.

The software architecture is open, and is based on the AIX/6000 operating system, IBM's implementation of the familiar and powerful UNIX. As a result, the SP2 can be incorporated into existing environments. The software architecture is also tuned to the SP2 hardware design for maximum usability and performance.

IBM provides a set of software products to support the SP2 system. This includes support for the following.

- System management and monitoring via the AIX Parallel System Support Programs
- Job management and scheduling via IBM LoadLeveler
- Message passing task communications via the AIX Parallel Environment and IBM AIX PVME, as well as other non-IBM offerings
- Connectivity via products such as Client IO/S and IBM Intermix
- Parallel application development and execution via the AIX Parallel Environment

In 1990, IBM announced the RISC System/6000 (RS/6000) family of concurrent superscalar workstations and servers, supporting clock rates ranging from 20 to 30 MHz.

Over the years, the POWER-based RS/6000 offerings have improved incrementally. Desktop, deskside, and rack system clock rates increased up to 62.5 MHz. More than ten of these models support a 32 KB I-cache. Additional compiler capability, especially in the area of restructuring data access patterns, has improved benchmarks and customers' code. Changes to the I/O area have increased Micro Channel bandwidth from 40 to 80 MB/S peak.

The RS/6000 systems are implementations of a Reduced Instruction Set Computer (RISC) architecture. As is characteristic of many RISC architectures, loads and stores provide the only storage access; arithmetic instructions use only register operands. Several instructions, often considered more complex than a traditional RISC definition, enhance performance. They include: (1) a floating-pint multiply-add (FMA) instruction; (2) a branch-on-count (BCT) operation; and (3) update forms of storage references.

The FMA compound instruction consists of a floating-point multiply and a dependent add. On POWER and POWER2 implementations, the FMA operation performs the multiply and add with a total latency of only two cycles. Independent FMA instructions can start every cycle. The FMA operation allows a peak MFLOPS rate equal to two times the MHz rate while using a single functional unit. Many experts credit the FMA instruction as a key component of the RS/6000's floating-point performance.

The BCT form of a conditional branch decrements and tests a special pur-pose register, the Count Register, to determine the outcome of the branch. Often a loop closing branch can be coded using the BCT form; the programmer loads the Count Register with an iteration count for the loop and the branch unit decrements and test this value independently of other Fixed-Point Unit (FXU) work. The RS/6000's BCT instruction and Count Register are examples of archi-tectural separation of resources that enhance the implementer's ability to exploit instruction-level parallelism. The FXU can off-load the loop count decrement and test operations, whereas the branch unit can accurately determine the fetch path without FXU synchronization.

The POWER2 processor complex consists of eight semicustom chips parti-tioned in a fashion similar to POWER: an Instruction Cache Unit (ICU), the FXU, the Floating Point Unit (FPU), four Data Cache Units (DCU), and a Storage Control Unit (SCU). The ICU prefetches instructions from the I-cache and places them in instruction buffers. ICU control logic decodes or analyzes the instructions in the buffers. The ICU executes ICU instructions (primarily branches), some-times affecting the prefetch path. The ICU dispatches non-ICU instructions to the FXU and FPU over a four-instruction-wide "instruction dispatch" bus (IBUS). The FXU and FPU process the respective arithmetic instructions. The FXU also processes storage reference instructions by generating and translating the ad-dresses before placing them on the cache address bus.

The FXU and FPU each contain two execution units. The memory, instruc-tion reload, and instruction dispatch buses, as well as the interface from the DCU to the FXU and FPU, are wider than in the POWER implementation to support additional execution units. The POWER2 implementation can execute six in-structions (branch, condition register, two fixed-point, and two floating-point) per cycle. Like POWER, the POWER2 FPU supports a compound FMA instruction that increases the peak execution rate to eight operations per cycle. The interfaces between the multichip module (MCM) and the memory and I/O units are compat-ible with those of POWER.

The POWER2 processor chip set offers two system configurations: a four-word memory bus with a 128-KB D-cache and an eight-word memory bus with a 256-KB D-cache. The I-cache is 32 KB for either configuration. A four-word system differs from an eight-word memory system in that it has only two memory cards and sup-ports transfers of four words per cycle between memory and the MCM.

The POWER2 I/O unit is the same as the one in the RS/6000 Models 580 and 980. The I/O unit implements the 64-bit Streaming Data protocol on the Micro Channel at 10 MHz. The I/O unit implements dual 64-byte buffers per DMA channel so that operations over the SIO bus and Micro Channel can full overlap. The I/O unit, along with some logic on the I/O planar, reduces the arbitration time on the Micro Channel from 400 to 100 ns. This improves bandwidth and bus uti-lization. In addition, enhancements to the protocol for the SIO bus include prefetch data commands from the I/O unit so that the DMA data from the memory are available to the I/O unit with minimum delay.

The switch provides the internal message passing fabric that connects all of the SP2 nodes together in a way that potentially allows all processors to send messages simultaneously. The hardware to support this connectivity consists of two basic elements, the switch board and the communications adapter. There is one adapter per node and one switch board unit per rack. The switch board unit contains eight logical switch chips, with 16 physical chips for reliability reasons and provides the connectivity of each of the nodes to the switch fabric as well as the rack-to-rack connectivity.

As a start, the switch fabric needs to be scalable from tens to thousands of nodes. In order to meet that objective a multistage network was chosen. The multistage network increases the amount of switch capability in a granular fashion as the number of processors grows. With this switch topology switch stages are added as the system grows to keep the amount of bandwidth available to each of the processors constant.

One measure of bandwidth that is useful for comparing machine designs is bisectional bandwidth. This is the most common measure of aggregate bandwidth for parallel machines and is loosely defined as follows: Define a plane that separates the parallel system into two parts containing an equal number of notes. This plane intersects some number of network links. Bisectional bandwidth is the total possible bandwidth crossing this plane through these links.

This term is often used to assess the scalability of topology. For crossbars, hypercubes, the SP2 Switch, and most multistage networks, bisectional bandwidth scales linearly with the number of nodes in the system. For a two-dimensional mesh bisectional bandwidth scales with the square root of the number of nodes. For a ring, bisectional bandwidth remains constant as nodes are added. Since the effective bandwidth per node for this measurement is the aggregate divided by the number of nodes, the mesh and ring provide reduced capability as the system grows. The SP2 system maintains constant bisectional bandwidth per processor independent of the size of the machine.

For fine-grain parallel applications, support for short messages with low latency and minimal message overhead is needed. A PIO short message capability was developed for this requirement: The processor can dump the message that it wants to send directly into the fabric and it goes to the other side. For long messages a DMA capability is needed: The processor can set up the transfer, go back to work, and then the message will be transferred.

Another critical characteristic of the system was multiuser support. A system that has a large number of nodes should be able to be subdivided at times to support many users. To support multiple users and because the low latency fine grain requirement demands that messages must come from user space (rather than with kernel calls) hardware protection between partitions or between jobs is required.

The protection is direct hardware protection, permitting latencies to decrease over time but keep the functional characteristics intact.

Another requirement for a multiuser system is fairness of message delivery. That is, one job can not monopolize the fabric. One needs to have a multiuser system with multiple jobs simultaneously running and all guaranteed to make progress. For that reason IBM picked a Packet Switched network.

Message flow control methods are commonly divided into circuit-switched and packet-switched methods. In circuit-switching flow-control, messages traverse a previously configured (and reserved) circuit in the network. A control packet or packets completes the configuration. The circuit remains reserved for transmitting message packets until it is torn down (unreserved) by another control packet. In some implementations the tail of a message packet unreserves the unit.

Circuit switch systems have two characteristics that are found undesirable. First, the circuit is reserved through the entire network for the duration of the message. This uses up more of the communication bandwidth than is necessary. Second, and more important, the circuit switch conflict resolution mechanism is to have the node losing the arbitration wait some period of time before retrying the connection. This has the potential of starving a particular node in a high traffic environment and was deemed absolutely unacceptable.

In packet-switching protocols, packets are self-routing. Each message packet contains a header that incorporates the routing information, and each switch element interprets this routing information to select the proper output port for that switch. Packet-switching techniques can be further characterized based on when and how the packets are forwarded between switches. Many distributed networks use a method known as store-and-forward, in which a switch element receives the entire packet before attempting to forward the packet to the next switch on the path to the destination node. Store-and-forward packet latency is therefore dependent on the packet length multiplied by the path length.

Cut-through packet-switching methods were devised to mitigate the effect of path length on total packet latency. In cut-through networks, the switch element examines the packet header's route information to determine the proper output port and immediately forwards the packet if the output port is not busy. In the event that that output port is busy, the packet is blocked and must be buffered in some fashion.

Cut-through methods can be further classified according to the selected method of packet buffering. Virtual cut-through methods require each switch element to either forward or buffer the entire packet. Thus, like store-and-forward, virtual cut-through requires switches to contain buffers large enough to store at least one packet, and flow control is packet based.

Wormhole routing is a more common form of cut-through, and is the method chosen for the SP2 Switch. Unlike virtual cut-through, switch elements utilizing wormhole routing need not buffer an entire packet when that packet cannot proceed. Instead, buffering occurs on subpacket basis. For instance, in the SP2 Switch, buffering occurs on a byte basis. As a result, blocked packets may be stored across multiple switch elements if the switch element that received the blocked packet header cannot store the entire packet in its own buffers.

Wormhole routing switch elements commonly contain buffers at each input port to minimize the number of switches required to completely buffer a blocked packet, thereby minimizing the number of communication links reserved for that packet. In addition, the SP2 switch element contains a unique central buffer called the central queue that dynamically allocates space for blocked packet bytes from any input port. The use of a central buffer allows an input port receiving more than its share of blocked packets to utilize more than its share of total buffer space inside the switch element. To highlight the use of the central queue, IBM termed their flow control method buffered wormhole routing.

To summarize, the SP2 network is a multistage, omega, buffered-wormhole routing packet-switch. If a message is traveling from node A to node B and another message needs to intersect it to go through from processors C to D, the messages can cut through each other and share the fabric. There is no blockage of one message by another message for long periods of time. If there is congestion on an outgoing path there is buffering on each of the chips such that messages are buffered and queued within the switch chip for fair delivery in a round-robin fashion, packet by packet.

Ease of job scheduling placed another requirement of the switch fabric: a uniform topology. The SP2 switch topology is uniform, which is to say that the fabric as an omega network has equidistant message traffic from any particular point in the fabric to any other point in the fabric. Algorithms do not place requirements on node selection or topology since all nodes are equidistant. This means that the scheduler does not need to worry about the physical location of the specific nodes selected for a job. I/O has the same flexibility: I/O server nodes can be located anywhere on the fabric.

To obtain an efficient message-passing implementation IBM designed "optimistic" protocols where the data are sent assuming that they will get there and there will be buffers. Rendezvous protocols also exist for very long messages in which an application can reserve space and then send the message at a later time.

The fabric must be reliable as viewed by the application. That is, the application user should not need to worry about retransmitting messages. The architecture provides end-to-end message protection. Checking codes are placed on messages that are used by the receiving node to check for errors.

An adapter-to-adapter acknowledge protocol architecture is used, using positive acknowledgment: If a message is sent from A to B the acknowledgment that it has been received is sent back from B to A. Until the acknowledgment is received, the communication subsystem owns the location of the message.

The multistage fabric can also provide redundant paths between nodes. This allows faulty components to be eliminated from the resource list for the route generator, and routes to be generated even when they are faulty components in the system.

The AIX Parallel Environment is used to create and execute parallel programs through an AIXwindows interface. This application development environment is made up of the following.

- A Parallel Message Passing Library (MPL) containing APIs for communications between executing tasks for a FORTRAN or C parallel program
- A Parallel Operating Environment (POE) for managing the development and execution of parallel applications
- A user friendly Visualization Tool (VT) for viewing the unique performance characteristics of a parallel application through trace driven program visualization through program and system statistics gathered during execution and late playback; and an online performance monitor to study operational activity of SP2 in real time
- A Parallel Debugger based on AIX debugging capabilities for parallel application task debugging

Parallel Operating Environment (POE) eases the transition from serial to parallel processing by allowing the use of standard AIX tools and techniques. This includes the areas of program compilation, scheduling, execution, and monitoring in a manner familiar to the UNIX programmer.

POE has an AIX Windows/6000 Desktop interface, Parallel Desktop, which simplifies use of POE. Source files, executables, and AIX Parallel Environment tools are represented on the Parallel Desktop as icons, easily manipulated to compile and execute parallel programs or start other AIX Parallel Environment tools. The Partition Manager of POE allows a user to compile his or her program using the XL FORTRAN or C Set ++ (C and C++) compilers and link in all required run-time libraries. It also controls parallel program execution by allocating nodes required to run the job, copies the executable tasks into the individual allocated nodes and invokes their execution, monitors the status of task execution, and communicates with job scheduling systems.

POE also contains two X Window System analysis tools: Program Marker Array and System Status Array. The Program Marker Array is a run-time analysis tool allowing the monitoring of a program's execution by presenting a graphical representation of the program tasks. This window representation consists of a number of small squares called lights that change color under program control. Each parallel task has its own row of lights. Subroutine calls inserted into these tasks change the light colors. This visual feedback can be invaluable in detecting application problems.

The System Status Array is an on-line analysis tool for examining the utilization of processor nodes. A window consisting of a number of colored squares representing SP2 or RISC System/6000 cluster processor nodes shows the percent of CPU utilization so the nodes that are least busy may be better utilized.

For identifying problem areas in parallel programs and to assist in tuning them, the Parallel Operating Environment contains the familiar AIX profilers, prof and gprof, allowing the determination for each executed task of

- Execution time in each routine;
- Number of times the routine is called; and
- Average milliseconds per call.

Visualization Tool (VT) is designed to graphically show the performance and communications characteristics of an application program (Trace Visualization), and also to act as an on-line monitor (Performance Monitoring). Essentially, the VT is a group of displays or visuals that each show some unique execution characteristic of an application or system. The VT conforms to Motif and X Window System standards.

During Trace Visualization, you play back program and system statistical and event records (trace records) generated during execution. These trace records are stored during execution into a trace file that, when played back with the Visualization Tool, lets you visualize information about the program as well as its use of the underlying system. For playing back trace files, the Visualization Tool provides an easy-to-use control panel that mimics the familiar operation of a compact disk (CD) player. Using the CD-like controls, which includes buttons for play, fast forward, rewind, and so on, you can easily control playback of the trace files. The information can help you analyze the program, tune it, optimize its use of the underlying system, and perform effective load balancing. The Visualization Tool is easy to understand and use. A few of the possible views follow.

- Connectivity graph
- Interprocessor communication
- Kiviat diagrams
- Message status
- Message status

During Performance Monitoring, the VT displays show current system activity at a configurable sampling frequency in realtime. This job-independent system-wide view indicates the operational status of each of the nodes. Kernel statistics describing system and communication subsystem activity on each node can be visualized from the user's workstation.

The AIX Parallel Environment builds on the familiar AIX dbx debugging facility but adds additional functions specific to parallel program debugging. The Parallel Debugger provides a command line interface, pdbx, and a graphical user interface, xpdbx designed for users not familiar with dbx. xpdbx uses a Motif-based X Window System interface and does not require the user to know dbx subcommands. Both interfaces allow the capability of grouping parallel tasks for debugging freeing the user from entering the same commands over and over again for each task.

The Parallel Debugger also lets you do the following.

- "Unhook" program processes not to be monitored.
- Debug object files.
- Set breakpoints at selected statements or run a program line-by-line.
- Debug a program using symbolic values, displaying them in the correct format.

AIX Parallel System Support Programs (AIX PSSP) are a collection of administrative software applications designed to provide the unique system management functions required to manage the Scalable POWERparallel 2 (SP2) system as a full function parallel system. AIX PSSP provides extensions to the AIX and contains all the software needed to install, operate, and maintain an SP2 system from a single point, the control workstation. Multiple components are included that are integrated together into a single software program product. In addition, a number of useful publicly available software packages are distributed with AIX PSSP.

5.4   KENDALL SQUARE RESEARCH—KSR1

Kendall Square Research (Waltham, MA) invested 6 years in developing its massively parallel computer system, the KSR1. Based on a novel architecture that retains the conventional shared-memory programming model, the KSR1 masks the underlying complexity of MPP (through a patented system called ALLCACHE) and allows programmers to port old applications or develop new ones.

By starting with a modest degree of parallelism (e.g., 32 robust processing units called APRD Cells), and continuing upward to as many as 1088 cells, the KSR1 can address a large number of specific, identified applications.

Kendall Square Research's MPP product family, the KSR1, has 36 systems in configurations ranging from 8 to 1088 APRD cells, the number of cells being designated by its configuration number. Table 5-2 shows eight representative configurations. The models scale linearly in all major dimensions of performance and capacity. List prices for KSR1 computer systems, including base disk capacity and peripherals, ranged from $500,000 for the KSR1-8 to over $30 million for the KSR1-1088.

TABLE 5-2

KENDALL SQUARE RESEARCH PRODUCT LINE					
Theoretical Peak Performance			Memory (Mbytes)	Maximum Disk Capacity (Gbytes)	Maximum I/O Capacity (Mbytes/sec)
Model	MIPS	MFLOPS			
KSR1-8	160	320	256	210	210
KSR1-16	320	640	512	450	450
KSR1-32	640	1280	1024	450	450
KSR1-64	1280	2560	2048	900	900
KSR1-128	2560	5120	4096	1800	1800
KSR1-256	5120	10,240	8192	3600	3600
KSR1-512	10,240	20,480	16,364	7200	7200
KSR1-1088	21,760	43,520	34,816	15,300	15,300

Source: Kendall Square Research, April 1992.

Each KSR computer system consists of one or more of the following.

- Processing Modules, each containing up to 32 APRD Cells including 1 Gbyte of ALLCACHE memory
- Disk Modules, each containing 10 Gbytes of formatted disk storage using RAID techniques
- I/O adaptors
- Power Modules, including an integrated battery-backup system for the entire machine, including disks

The KSR1 was designed so that if a critical component fails it may be bypassed, and in certain conditions, replaced while the system remains online. The KSR1's modular construction allowed a customer to add capacity as needed.

Each APRD (ALLCACHE Processor, Router, and Directory) Cell contains a total of 12 KSR-designed CMOS devices of six types.

1. A Floating Point Unit that operates on 64-bit IEEE floating-point numbers.

2. An Integer Processing Unit that operates on 64-bit integers.

3. A Cell Execution Unit that generates addresses and executes logical and arithmetic operations on 40-bit addresses.

4. An External I/O Unit that moves data between peripheral devices and the APRD Cell.

5. Four Cache Control Units.

6. Four Cell Interconnection Units.

In addition, the APRD Cell contains 32 Mbytes of Local Cache memory and 512 Kbytes of subcache memory.

The KSR1 processor implements a 40-bit address space, uses two 64-bit buses for transport of data and instructions, and can perform arithmetic operations on IEEE-standard 64-bit floating-point numbers at a peak rate of 40 MFLOPS. To reduce memory traffic, each APRD Cell has large register sets: 64 floating-point registers (64-bit); 32 integer registers (64-bit); and 32 address registers (40-bit).

The KSR-designed CMOS devices are implemented in 1.2 micron lithography and the APRD Cells are packaged in an 8 × 13 × 1 inch circuit-board assembly.

ALLCACHE automatically moves an address set requested by a processor to the 32 Mbyte Local Cache associated with that processor. ALLCACHE exploits a property of address-reference sequences known as "locality of reference." Programs and data, once referenced, are likely to be referenced again. ALLCACHE keeps memory traffic close to the processor that is using the data, which reduces latency and communications congestion.

The Search Engine is responsible for finding addresses and their contents and relocating them to a processor's Local Cache when referenced by that processor, maintaining sequential consistency among Local Caches. The address and data remain at the Local Cache until space is needed by another address. The search engine interconnects and provides routing and directory services for a collection of Local Caches. As a result, the collection of Local Caches behaves as a single shared-address space.

The KSR1's operating system, KSR OS, is an enhanced version of OSF/1 UNIX. It runs symmetrically on all processors and provides flexibility in allocating computational work among multiple processors. The scalable general-purpose environment simultaneously supports multiple computing modes, including batch, interactive, OLTP, and database management and inquiry running concurrently. KSR OS supports the most widely accepted UNIX interface standards, AT&T SVr3, BSD 4.3, Posix 1003.1, XPG, and FIPS 151.1.

KSR1 programming languages are Fortran (with automatic parallelization) and ANSI C. Parallel Programming can be fully automatic, semiautomatic, or manual. The parallel programming environment is based on KSR's PRESTO parallel runtime system that dynamically executes runtime decisions based on compiler-generated or programmer-specified directives, available processing resources, and priorities established by systems administrators. For example, PRESTO dynamically adjusts to the number of available processors and the size of the current problem and dynamically optimizes data locality.

The KSR1 was designed to operate within multivendor environments and includes interconnections through NFS, Ethernet, X.25, HDLC, and SNA. For support of third-party peripheral devices, the KSR1 provides SCSI and VME interfaces.

ALLCACHE memory creates a familiar and easy-to-use development environment that reduces programming difficulties associated with highly parallel processing systems. ALLCACHE automates the addressing and location of memory. Users can port existing application programs to the KSR1, and develop new applications using industry standard development environments.

ALLCACHE is the first memory architecture to deliver the conventional, sequentially consistent shared memory programming model in a highly parallel computer. ALLCACHE combines this memory model, used by traditional supercomputers and mainframes, with the scalability of highly parallel systems. Scalability allows users to add computer resources in incremental and cost-effective steps, without changes in software and without performance degradation. Systems that implement sequential consistency guarantee that a program behaves in the most intuitive manner to the programmer: the result of a program execution on parallel processor, which carries out tasks in a sequential fashion.

ALLCACHE is, literally, all cache. Thirty-two MBytes of cache memory as attached to each KSR1 processor. That is the only memory there is. No permanent physical location exists for an "address," as illustrated in Figure 5.4-1. Rather, addresses are distributed and shared based on processor need and access patterns. To the application developer, memory is a single resource. Although Kendall Square Research designed the KSR1 architecture to support a 64-bit address space, the current KSR1 implementation supports a 40-bit address space—providing one Terabyte (one trillion bytes) of virtual address space per process. This greatly exceeds the memory capacity required in today's high performance computing environments.

When addresses are requested by a processor through the execution of a load, store, or branch instruction, ALLCACHE automatically moves them to the processor's physical "neighborhood" (a 32-Mbyte cache memory module).

The KSR1 memory architecture consists of two levels of address space: Context Address (CA) space and System Virtual Address (SVA) space. CA space is the programmer's interface to memory. There are many CA spaces in a system. SVA space stores the data from all context address spaces. There is only one SVA space. Load, store, and branch instructions generate CAs, which are translated by the processor into SVAs.

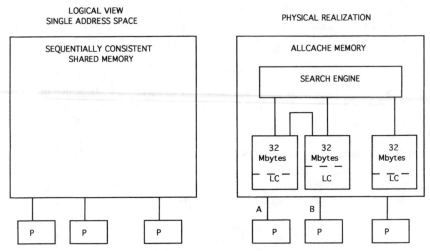

P = 64-BIT SUPERSCALAR; LC = LOCAL CACHE.

ALLCACHE: DATA MOVES TO THE POINT OF REFERENCE ON DEMAND. THERE IS NO
FIXED PHYSICAL LOCATION FOR AN "ADDRESS" WITH ALLCACHE MEMORY.

Figure 5.4-1.   ALLCACHE Memory System.

To coordinate this high level of memory interaction, the KSR1 employs a
Search Engine. Attached to each KSR1 processor, the Search Engine is a distributed mechanism responsible for efficiently finding addresses and their contents,
relocating them to a processor's local cache when referenced by that local processor, and maintaining sequential consistency between caches. The address and data
remain at the local cache until the space is needed by another address.

The Search Engine interconnects local caches and provides routing and directory services for the collection of local caches, while maintaining coherence
throughout the system. As a result, the collection of local caches behave as a single shared address space.

However, providing a shared memory architecture on a parallel computing
platform is not adequate by itself. The challenge was to build a shared memory
model that allows the system to scale to large sizes. The KSR1's scalable interconnection system is a two-tier hierarchy of "search groups" composed of clusters of processors and memory.

The first level, called Search Group:0, consists of clusters of up to 32 processors. The second level, called Search Group:1, consists of up to 34; Search
Group:0s, or 1088 processors.

The ALLCACHE design eliminates the traditional parallel bottlenecks in
memory traffic, provides a simple programming model, and supports an entire

environment of familiar software tools. Users can preserve traditional programming approaches, while utilizing the price/performance advantage of highly parallel systems to build the next generation of high performance computing solutions.

At the heart of the ALLCACHE memory architecture is the KSR1 ALLCACHE Processor, Routing, and Directory cell (APRD cell), which includes a 64-bit superscalar processor, 32 MBytes of cache memory, and a portion of the Search Engine.

The APRD cell is composed of six types of full custom complementary metal-oxide semiconductor (CMOS) chips, packaged on a 20 × 32 centimeter (8 × 13-inch) circuit board, as noted in Figure 5.4-2. Using 1.2-μm CMOS, each KSR1 custom chip contains up to 450,000 transistors. The processor is clocked at 20 MHz and executes two instructions per cycle.

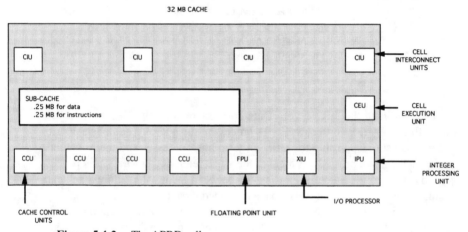

Figure 5.4-2.   The APRD cell.

5.5   CRAY T3D

The Cray T3D system incorporates hundreds or thousands of DEC Alpha microprocessors and supporting communications and synchronization logic into a high-speed, three-dimensional torus communication network [1]. The memory is physically distributed; each processing element combines a processor with its own local memory. The memory is also globally addressable in that any processor can address any word of memory on any other processing element.

MPPs (massively parallel processors) use a nonuniform memory architecture. Processors can access their local memory faster than memories on other processing elements. On the Cray T3D system, this delay in accessing remote memories is small when compared with the delays found on other MPPs, but the delay is still significant.

On many MPPs, communicating with distant processors is significantly slower than communicating with neighboring processors. The three-dimensional torus shape of the Cray T3D system lessens this effect significantly. In effect, there are no far neighbors in a three-dimensional torus network.

Multiple mechanisms are provided within the Cray T3D system for a processor to access the memories of the other processors. These mechanisms allow the Cray Research software to hide the delays inherent in this remote communication. A processor can request data from a remote memory before they are needed, and then continue with useful work until the data arrive. This prevents the processor from stalling while waiting for the remote data. This feature is termed prefetch.

The Cray T3D system uses technologies developed for its Cray Y-MP C90 class machines to clock the communications and synchronization networks at the same high speed as the DEC Alpha processors (150 MHz). This allows applications to make remote memory requests more frequently and efficiently.

The operating system is distributed between the Cray T3D system and a UNIX Server. Any of the multiprocessor Cray Y-MP C90 systems or model E Cray Y-MP systems can be used as the UNIX Server. In this section the terms, UNIX Server, Cray Y-MP C90, and Cray Y-MP are interchangeable.

Cray Research's MPP system is heterogeneous, combining the strengths of a Cray Y-MP C90 system with the Cray T3D system. One or more high-speed channels can be used to communicate between these systems. This heterogeneous architecture has several major benefits.

- All the I/O functionality of the Cray Y-MP C90 hardware and software is available on the MPP. This includes multiterabyte file systems, high-speed disks, HIPPI and FDDI network interfaces, 3490 tapes, and tape silos. This also includes the extensive UNICOS operating system capabilities.

- This heterogeneous system allows the job mix to be split between the parallel-vector resource and the MPP resource. Some applications might run exclusively on the general-purpose Cray Y-MP C90 system, whereas others run exclusively on the Cray T3D system.

- Individual applications also can be distributed between these two resources, with the appropriate portions of applications running in parallel on the appropriate architectures. This provides an evolutionary path to move portions of the application from the Cray Y-MP C90 system to the

Cray T3D system, as new parallel algorithm and computational methods arise.

The result of this heterogeneous architecture is the fastest general-purpose high-end computing system available, capable of running the broadest range of MPP applications.

The Cray T3D operating system is called MAX, which is an acronym for *m*assively *p*arallel *U*NIX. It is a microkernel-based distributed MPP operating system.

Each processor within the Cray T3D system runs a microkernel. This microkernel provides basic operating system functionality, such as interprocessor communication and memory management. The size of the microkernel is held to a minimum, with most UNIX functionality placed elsewhere in the distributed operating system. When a system request is made for a function not resident in the microkernel, the microkernel uses fast interprocessor communication mechanisms to forward the request to a UNIX Agent that runs on the Cray Y-MP C90 server.

The UNIX Agent runs as a user process on the Cray Y-MP C90 server and therefore has access to all the functionality of Cray Research's UNICOS operating system. This includes all the Cray Y-MP I/O hardware described in the preceding, as well as a set of UNICOS features. These features include batch queuing systems, fast asynchronous I/O, security, distributed file systems, an extensive set of networking protocols, and other UNICOS features.

To access the Cray T3D system, a user logs on to the UNIX Server and uses the UNICOS program development tools to prepare a Cray T3D application. The editors, compilers, debuggers, source code browsers, performance analysis tools, and other tools run on the UNIX server.

The MPP application executes on the Cray T3D system, but the user environment on the UNIX Server remains intact in the full UNICOS environment. For example, the Cray T3D process remains connected to *standard in* and *standard out*. The Cray T3D process reads and writes the same files in the same file system as the UNIX Server.

The MAX operating system is a multiuser system. The hundreds or thousands of MPP processors on the Cray T3D system can be divided into partitions, allowing many applications to use the MPP at the same time. A user can request that a specific number of processors be allocated to a Cray T3D application. At run time, the operating system selects a set of processors to meet this request and organizes them into a partition. For example, on a 512-processor Cray T3D system, a user might request only 128 processors for a particular application. The operating system would then reserve one-fourth of the Cray T3D processors for that application.

The system administrator can define a set of processors to be a *pool*. Characteristics can be assigned to this pool, such as whether the pool is intended for batch use, interactive use, or both. The administrator can also assign time limits and other execution attributes necessary to control the Cray T3D resources in a production environment.

At run time, a program requests a *partition* from within a *pool*.

The MAX system implements *space sharing* to allow multiple users to use the Cray T3D system at the same time. Multiple applications run on the machine in multiple partitions. Each application starts in a partition and has that partition all to itself until it finishes. The operating system will not interrupt a partition to run another application until the application in the partition finishes.

I/O on the Cray T3D system supplies the full functionality of UNICOS I/O, with the MAX operating system. All the disk, tape, and networking devices are supported, including HIPPI disk arrays, D2 and 3490 tapes and tape silos, and high-speed networks, such as HIPPI and FDDI.

It provides a shared file system between the UNIX Server and the Cray T3D system. Both machines access the same physical files, on the same physical disks, controlled by the same file system server.

All I/O data and control go through the UNIX Server. All the I/O clusters are physically connected to the UNIX Server and all the I/O therefore passes through the UNIX Server. For example, a Cray T3D program issues a read disk system call. The microkernel sends this request to the UNIX Agent associated with that program's partition. The UNIX Agent issues the UNICOS system call to perform the disk read. UNICOS asks the appropriate I/O cluster to send the data to the UNIX Agent. The UNIX Agent forwards the data over a high-speed channel to the Cray T3D system where it is routed to the processing element that made the read request.

The Cray T3D design includes redundant processing nodes to allow the hardware to continue functioning at full capacity even if one of the processors should fail. The software maps the physical three-dimensional layout of processors into a virtual three-dimensional layout. This allows the spare processors to be mapped into the virtual three-dimensional layout, anywhere in the physical three-dimensional network. Thus, with a software reconfiguration, the system administrator can replace a malfunctioning processor with a spare processor instead of waiting for an engineer to repair the hardware.

The Cray Research MPP programming environment provides tools that programmers can use to minimize communication overhead and maximize the performance of their applications. As described previously, most MPPs require careful programming to manage the communication necessary between the distributed memories and the individual processing elements. Cray Research's programming languages provide the additional constructs necessary to program this communication efficiently.

Cray Research's goal is to make this communication implicit in the algorithms, minimizing the burden on programmers. The Cray T3D programming models allow programmers to code the communication explicitly, when more control is needed. This mix of explicit and implicit communication techniques gives programmers the tools to write code with minimum effort, without sacrificing the ability to optimize the time-critical sections.

The Cray Research programming models provide upward compatibility from generation to generation of Cray Research MPPs. The microarchitecture of each generation may change, but the macroarchitecture will remain constant. If the microprocessor instruction set changes, the programming language will not change. The application would need to be recompiled, but no source code changes would be needed.

The Cray T3D system supports several programming styles: message passing, data parallel, and work sharing.

Message Passing. Message passing provides explicit communications only. In the message-passing model, programmers must explicitly send messages to request information from other processing elements. The Cray T3D message-passing library is based on the Parallel Virtual Machine (PVM 3.0) developed at the Oak Ridge National Laboratory. Cray Research views message passing as a key programming method on MPPs and is giving it full support, including special fast message passing hardware. Message-passing codes written for other MPPs will port to Cray Research's MPP system, and they will generally run faster on the Cray T3D system.

Cray Research's implementation of PVM supports Fortran, C, and C++ and can be used in conjunction with the HPF and MPP Fortran programming models described in the following.

The High Performance Fortran (HPF) model. The High Performance Fortran (HPF) model uses mostly implicit communications. Cray Research is a strong supporter of the emerging HPF programming standard and is an active participant in the HPF Forum. HPF uses Fortran 90 data-parallel constructs to allow programmers to avoid explicitly coding the communication. HPF uses Fortran-D-like data distribution constructs to divide the arrays among the distributed memories.

Cray Research's MPP Fortran model. Cray Research's MPP Fortran model provides a combination of implicit and explicit communications. MPP Fortran is closely related to HPF, with more options for optimizing the code when necessary. It allows programmers to write HPF-like code, using Fortran 90 data-parallel programming, and to avoid explicit communications. Programmers distribute data among the processors much as they would when using HPF. MPP Fortran includes enhancements that give programmers greater control of communication and synchronization, allowing them to code the communication explicitly, if necessary. In addition, MPP Fortran includes a work-sharing model, which allows work to be shared among the processors using standard FORTRAN 77 DO loops. The work-sharing model supports fine levels of granularity by distributing loop iterations across processors. It also supports dynamic distribution of work for load balancing.

Programmers can mix the data-parallel, work-sharing, and message-passing styles of programming within an application. They can use PVM message passing with MPP Fortran and with HPF. Because the Cray T3D system offers choices rather than dictating one particular style, users can decide which approach is best for their application. This variety of styles will also ease the porting of existing programs from other massively parallel systems and from Cray Research parallel-vector systems.

REFERENCE

1. Vendor literature, Cray T3D Software Overview Technical Note, Cray Research, Inc., Special Draft Edition for IEEE, Supercomputing '92; Cray Research Park, Eagan, MA.

5.6   PARSYTEC

Parsytec is the parallel processing market leader in Europe [1]. Parsytec has over 1200 customers in 22 countries. Parsytec has also secured a range of strategic collaborations.

- As "inventor" of the transputer INMOS Ltd. from Bristol, a company in the SGS-Thomson group, is a major partner.
- The Dutch software house of ACE bv in Amsterdam will develop compilers for Parsytec GC. ACE bv has played a leading role in the construction of compilers and development tools; for example, in 1976 they were the first to provide a European UNIX implementation. ACE is producing highly optimizing, validated compilers for Fortran, C, Pascal, and Modula for the Parsytec GC series, including numerous tools for testing and monitoring.
- Great impetus has been given to the Parsytec GC series by other collaborations in the design of software tools, peripheral equipment, and communications.

The Parsytec GC/PowerPlus combines the state-of-the-art RISC processor PowerPC™-601 and advanced European transputer communication technology to provide an outstanding compute resource for the following.

- Research in parallel processing
- Numerical simulations

- (Visual) pattern recognition
- (Real-time) image synthesis
- Optimization using discrete mathematics

GC/PowerPlus achieves top-end supercomputer performance with a scalable architecture supporting thousands of processors and offers a clear upgrade path to PowerPC-604 and PowerPC-620 microprocessors.

The GC/PowerPlus distributed memory architecture employs a scalable routing network and a node design featuring a scalable number of application and communication processors. The overall architecture is optimized for most efficient execution of parallel programs composed of compute and communication threads. The GC/PowerPlus architecture therefore supports parallelism at all three levels.

1. **Per-chip parallelism:** The PowerPC-601 executes three instructions per cycle, the T805 transputer supports eight independent concurrent communications.

2. **Per-node parallelism:** Four communication and two application processors form a single GC/PowerPlus node and allows six parallel threads.

3. **Per-application parallelism:** A large number of physical nodes (e.g., 1024) is connected through a bidirectional fat-grid network with packet message routing.

The node architecture uses a "variable balance architecture" (VBA) and supports a variable number of application and communication processors and memory modules.

The GC/PowerPlus 2+4 node design has been selected after extensive simulations as the best compromise between cost and scalability. The use of intelligent transputer communication processors has already triggered the development of advanced communication firmware to support monitoring, fault-tolerant communication, global operations, higher level protocols, and scatter/gather communication.

The two-dimensional fat-grid topology employs four links to connect to neighboring processors and uses all four links simultaneously to transfer messages. Large systems feature an additional redundant node for hot switch-in for each 16 standard nodes. This patented fault-tolerant implementation and Parsytec's parallel checkpoint/restart feature guarantee good system availability vital for critical production runs. Scalable I/O performance is achieved via the GC/PowerPlus boundary links to the parallel storage system.

The GC/PowerPlus runs the POSIX compliant parallel operating environment PARIX™. Maintaining UNIX compatibility ensures transfer of applications.

PARIX™ hides the implementation of both the network architecture by providing virtual thread-to-thread communication channels and the VBA node implementation by providing virtual processors. PARIX™ offers an easy to understand paradigm: An application allocates communicating virtual processors (typically many per processor). Exchanging data (includes control flow synchronization according to the CSP model) usually involves delays while other virtual processors continue execution on the same physical processor. The PARIX™ implementation has been optimized to support this technique called latency hiding. Efficiency then heavily depends on context switching time. PARIX™ achieves more than two orders of magnitude improvement over standard UNIX in process switching time (3 to 7 μs) and requires just 100 kB memory per node.

For program development, PARIX™ provides necessary tools for parallel program development, such as optimizing compilers for C, C++ and FORTRAN, debuggers, performance analyzers, parallel programming tools such as Forge-90 and multiuser administration software. PARIX™ is object code compatible to RS/6000. The range of additional parallel processing paradigms include Express, MPI, PVM, Linda, and P4.

The Parsytec parallel storage system connects to the boundary links of Parsytec GC/PowerPlus. Four or eight links (up to 12.8 MB/s) connect to 8 GB disk nodes (RAID optional). The PARIX™ parallel filing system PFS creates a UNIX SVR4 compliant parallel access model to distributed mass storage. Standard sequential accessibility over fast networks is maintained, for example, for visualization of large data sets. Multiple disk nodes can be added to enhance capacity and bandwidth. Parallel backup is implemented by autonomous DAT devices.

PARIX™ supports a variety of tools for software development including: C, C++, F77, and F90 compilers, performance monitors, debuggers, for example, GNU xgdb, X11, parallel programming interfaces such as MPI, PVM, P4, Linda, Express, libraries such as VPROC, VTOP, BLAS, sockets and the parallel filing system PFS.

Parsytec performance is described in Table 5-3. PARIX™ features are detailed in Table 5-4.

The Motorola PowerPC™ FORTRAN 77 compilation system provides a powerful set of software development tools designed specifically for the requirements of RISC microprocessors. The compilation system consists of a highly optimizing FORTRAN compiler and a complete set of libraries (see Table 5-5).

REFERENCE

1. Schmitz, Bob, Vendor literature and personal communication, Parsytec Inc., West Chicago, IL.

TABLE 5-3

PARSYTEC PERFORMANCE		
	Per Node	**128 Processors**
System	2 PowerPC™-601-80 4 IMS T805-30	128 PowerPC™-601-80 256 IMS T805-30 dis- tributed memory MPP

Operating System Performance
    64-bit IEEE floating point
    32-bit IEEE floating point
    32-bit integer
	PARIX™ 1.2
	80 MFLOPS
	160 MFLOPS
	160 MIPS
		PARIX™ 1.2
		10 GFLOPS
		20 GFLOPS
		20 GIPS

Memory Streaming caches bandwidth	16 ... 128 MB 160 MB/s sustained	1 ... 8 GB

Communication
 	
    Message setup time
    Minimum network latency
    Bisectional bandwidth
	16 OS links
	35 MB/s sustained
	80 MB/s peak
	5 μs
	40 μs
		1 GB/s sustained
		
		5 μs
		60 μs
		70 MB/s

Max I/O Bandwidth	35 MB/s	280 MB/s
Redundant Nodes		4 (eight processors)

5.7   CONVEX SPP-1000

The Convex Exemplar SPP-1000 system is a scalable multiprocessor computer system designed to provide supercomputing-class resources for a wide range of applications. The underlying system architecture provides performance through a combination of high-performance processors, a hybrid memory system, and a high-performance I/O subsystem. At the same time, use of off-the-shelf components, such as the RISC CPU, permits price/performance that tracks that of commodity microprocessors.

The Exemplar supports thousands of commercial and technical applications. In addition, the programming environment is designed to port and/or develop new applications on a MPP system. This environment is supported by

- An Application Binary Interface (ABI) compatible with Hewlett Packard's HPUX operating system;
- An underlying familiar shared-memory programming model; and
- A toolset designed for development and porting of MPP applications.

TABLE 5-4

PARIX™

Features
  NanoKernel implements nodes as abstract processor with:
    User space communication
    Reliable communication semantics (such as memory store/load)
    Multithreading with latency-hiding (3 μs)
    Multiprocessor nodes with local load-balancing
    Protected memory and channel address spaces
    Power ABI compliant system-call interface (subset)
  SunOS 4.x and Solaris 2 development environments appearing as one or multiple
    UNIX-nodes
  Support for SPMD (single program multiple data) programming
  Support for asynchronous (object-oriented) parallel programming
  Mapping of application-defined communication topologies onto hardware
  Multiuser operation with possible multinode protection domains
  Posix-compliant Parallel File System (PFS) with scalable bandwidth

Performance
  Thread switch 3 μs (7 μs with touched FPU registers) PowerPC 601-80
  Overhead-free (user space) asynchronous communication
  Virtual channel processors and computing processors (PowerPC) work concurrently. Intelligent communication protocols implemented by node-firmware supported
  35 MByte/s sustained communication bandwidth per node using the virtual channel protocol; GC/PowerPlus with intelligent T805 communication processor

Compliance
  Posix 1003.x and XPG/3
  FORTRAN 77, K&R C, ANSI-C
  FORTRAN 90, HPF, C++ (tba)
  PVM, P4, Linda, Sun-RPC, X11R4, MPI (tba)
  Numerical Libraries: BLAS

Tools
  Source level debugging (xgdb, DETOP parallel debugger)
  Performance and program monitoring (PATOP)
  Forge90

Platforms Supported
  Parsytec GC/PowerPlus, GC/T9, and GCel
  Parsytec PowerXplorer and Parsytec Xplorer
  Parsytec MultiCluster

PARIX™—Features
  Virtual topologies
  Virtual processors
  Multithreaded programming
  Direct thread-to-thread messages
  Synchronous and asynchronous communication
  Mailboxes
  Dynamic code loading
  Transparent host access
  Checkpoint/restart
  POSIX parallel filing system

TABLE 5-5

FORTRAN COMPILATION SYSTEM
Features 　ANSI-compliant FORTRAN front end 　VAX/VMS language extensions 　PowerOpen ABI compliance 　Assembly language and XCOFF code generation 　Integration with Motorola timing and architectural simulators 　Industry-leading performance
PowerPC Compliance 　The compiler is tuned to make full use of PowerPC architectural features, 　including: 　　Superscalar instruction scheduling 　　Efficient register allocation, including support for renaming 　　Branch prediction and branch acceleration support 　　Effective use of cache and memory hierarchy 　　Specific optimizations for 601, 603, 604, and 620 microprocessors
Performance 　The global optimizer implements a range of optimizations, including: 　　Local and global common subexpression elimination 　　Priority-based graph coloring register allocation 　　Software pipelining 　　Loop unrolling 　　Constant and copy propagation 　　Aggressive load/store removal 　　Interprocedural analysis and optimization 　　Function inlining capability 　　Execution profile feedback mechanism

The system is designed to provide the benefits of a scalable parallel processing system while maintaining a familiar programming model. This programming model allows the developer, compilers, and applications to view the system as a number of processors sharing a large physical memory and a number of high-bandwidth I/O ports. Logically, the system appears as a 4- to 128-processor MIMD (multiple instruction multiple data) shared memory system, as shown in Figure 5.7-1.

Underlying this view are sophisticated hardware and software that manage a globally shared distributed virtual memory (GSDVM) system. The GSDVM provides the scalability benefits of a distributed memory while maintaining the development environment of a shared memory system. In addition, the GSDVM automatically maintains complete data coherency throughout the system. The GSDVM dynamically distributes data throughout the memory system, which in-

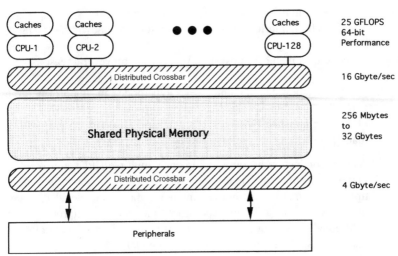

Figure 5.7-1.   A user's view of the Exemplar SPP-1000 hardware.

creases application performance by eliminating excessive interhypernode data movement.

The Exemplar SPP-1000 system architecture is based on commodity RISC processors, which derive the majority of their required data bandwidth from local instruction and data caches. These caches (unlike the ALLCACHE design of the KSR-1) are generally small with respect to the size of an MPP application, therefore, it is very important to minimize the cache fault refill time to maintain high performance. In a small (i.e., 4- to 8-processor) multiprocessor system, all processors can share a common high bandwidth interconnect to reach memory within a reasonable latency. This interconnect can also provide cache coherent access to shared data in a simple way.

Systems with hundreds or thousands of processors must rely on some form of hierarchical memory structure to balance the simultaneous requirements of low latency and high bandwidth. These systems provide massive bandwidth by distributing the memory structure and providing simultaneous access to each segment of each memory element in the hierarchy. Latency is minimized through a hierarchy of caches that can transparently move the data closer to the processor, and/or by dynamically copying the required data to memory elements that are closer to a particular processor.

The SPP-1000 supports a two-level hierarchy of memory, each of which is optimized for a particular class of data sharing. The first level consists of a single-dimensional torus of identical processor/memory hypernodes. Convex also expands the normal concept of a toroidal interconnect by splitting each link of the torus into four interleaved links for additional bandwidth and fault resilience.

This hierarchical memory subsystem was chosen for several reasons.

1. The low-latency shared memory of a hypernode can support fine-grained parallelism within applications, thus improving performance. This is often accomplished by simply recompiling the program with the automatic parallelizing compilers.

2. As semiconductors become more dense, multiple CPUs are likely to be placed on a single die. Thus, the first level of the hierarchy will be multiple CPUs sharing a common memory; the second level of memory hierarchy will be "off-chip" (possibly to a nearby multiprocessor chip). Additionally, as multiprocessor workstations are introduced, clusters of these workstations will follow the same model.

3. This system organization is a superset of a cluster of workstations, traditional experimental MPP systems, and SMP systems. Processors within a hypernode are tightly coupled to support fine grained parallelism. Hypernodes implement coarser grained parallelism with communication through shared memory and/or message passing mechanisms.

The fundamental idea behind the Convex Exemplar architecture is that a hypernode is a shared memory multiprocessor. An Exemplar SPP-1000 is a group of hypernodes sharing a common, coherent distributed memory and I/O subsystem.

The hypernodes each contain their own 5×5 memory crossbar; very similar to traditional supercomputer memory subsystems. The crossbar provides high bandwidth, low latency, nonblocking access from CPUs and I/O channels to the hypernode-local memory. The crossbar is implemented in high density Gallium Arsenide (GaAs) gate arrays. As in earlier Convex systems, the use of GaAs permits high performance while keeping the facilities requirements to a minimum.

The use of a crossbar implementation prevents the performance drop-off associated with systems that employ a system-wide bus to handle CPU and I/O traffic. Figure 5.7-2 illustrates the physical components of an Exemplar hypernode.

A hypernode supports 256, 512, 1024, or 2048 MBytes of 60ns DRAM, physically distributed on four boards. There are two physical banks per board. Each bank can transfer a 64-byte cache line (eight longwords) in eight clock cycles (plus startup time, which will depend on the location of the requester).

Note that the Convex Exemplar provides hardware support for global shared memory access, whereas a true distributed memory machine can only emulate shared memory by moving pages from node to node under software control. Cache lines may be automatically copied and encached between hypernodes

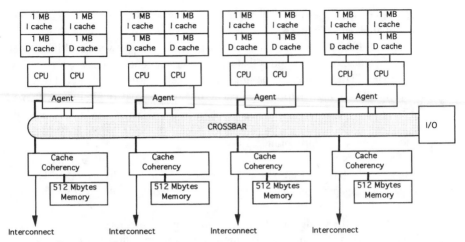

Figure 5.7-2.   An SPP-1000 hypernode.

without software intervention, resulting in lower overhead to programs using shared memory.

Multiple hypernodes are interconnected with a low latency interconnect called CTI—Coherent Toroidal Interconnect. CTI is a Convex implementation of the IEEE standard 1596-1992, SCI (Scalable Coherency Interface). This interconnect combines high-bandwidth with low-latency to provide system-wide coherent access to shared memory.

The CTI also supports messaging protocols. Messages are used by the operating system and distributed memory applications for inter-hypernode communication.

The CTI is composed of four unidirectional rings attached to each hypernode. Four rings are used to provide higher interconnection bandwidth, lower inter-hypernode latency, and redundancy in case of ring failure.

All CTI communication is done in packets. Packets contain a 16 bit destination hypernode identifier, as well as a 16-bit source hypernode identifier, and a 48-bit address to be accessed within the destination hypernode. The base protocol supports guaranteed forward progress, guaranteed delivery, fairness, and basic error detection and recovery.

To minimize memory request latencies across the CTI, each hypernode contains a cache of memory references made over the interconnect to other hypernodes. This is referred to as the *CTIcache*. Any data that have been moved into a CPU cache on the same hypernode, and are still resident in the CPU cache, are guaranteed also to be encached in the CTIcache. Consequently, the CTIcache directory information can be used to locate any global data that are currently encached by the hypernode. The CTIcache is physically indexed and tagged with the global physical address.

The SPP-1000 system guarantees cache coherence between multiple hypernodes; two or more hypernodes that map the same global address will get a consistent view. This is done by maintaining a linked sharing list that contains a list of all the hypernodes sharing each cache line, or the hypernode the exclusively owns the cache line. Within every hypernode a record is kept of which CPUs have encached each line in the CTIcache, so that interconnect coherency requests can be forwarded to the appropriate CPUs.

Developers may use special instructions in the SPP-1000 architecture that prefetch cache blocks from remote hypernodes to the local hypernode CTIcache. This accelerates performance by masking the interconnect latency from the application.

Note that the combination of hypernode shared memory and the CTIcache effectively eliminates much of the internode communication that is inherent in a traditional distributed memory architecture. Thus, measurements usually reserved for such machines, such as bisection bandwidth, are normally not applicable to this architecture.

The CPUs in the SPP-1000 systems are Hewlett Packard's seventh generation PA RISC microprocessor, the model PA7100. It is a single-chip CMOS implementation containing 550,000 transistors using .8-μm technology.

The processor chip connects directly to an external 64-bit-wide instruction cache and an external 64-bit-wide data cache. These caches are direct mapped and are composed of industry standard high speed SRAMs that are cycled at the processor frequency. At 100 Mhz, each cache has a read bandwidth of 800 MBytes/second and a write bandwidth of 400 MBytes/second.

The use of off-chip caches allows a larger primary cache, increasing the hit rate and lowering the average memory access time. In addition, processors that have a smaller on-chip cache usually require a secondary cache that is not clocked at the processor frequency. The result is a substantial penalty for a first-level cache miss.

The CPU has a unified instruction/data translation lookaside buffer (TLB). The TLB is organized as 120 fully associative page entries. A page entry is 4 KB in size. Besides the 120 page entries, the TLB contains 16 block entries. Each block entry is capable of mapping contiguous virtual address space ranging in size from 128 to 16,384 pages.

The floating point unit (FPU) is integrated onto the CPU chip, and therefore is clocked at the same speed as the CPU. The FPU implements IEEE 754 compliant single and double precision math.

The floating point functional units include multiply (both floating and integer), divide/square root (running at 2X the CPU clock frequency), and add/logical, which performs addition, subtraction, and formation conversions. The functional units are interfaced to a register file that has 28 64-bit registers, each of which can be used as two 32-bit registers for single precision operations. The register file has five read ports and three write ports to allow concurrent execution of a multiply, add, and a load or store.

In addition, the floating point unit supports compound instructions that utilize the add/logical and multiply units simultaneously. These instructions perform multiply and add combination or a multiply and subtract combination.

The CPU has a six-stage pipeline. These six stages are

1. Instruction Cache Read (IR);

2. Operand Read (OR);

3. Execute/Data Cache Read (DR);

4. Data Cache Read Complete (DRC);

5. Register Write (RW); and

6. Data Cache Write (DW).

All LOAD instructions execute in a single cycle and require only one cycle of data cache bandwidth. Since the instruction and data caches are accessed on separate busses, there is never any conflict between data cache access and instruction fetches.

The processor can execute one integer instruction and one floating point operation in the same clock cycle. Here, "integer instruction" includes loads and stores of floating point registers, and "floating point operation" includes the compound floating multiply-add and floating multiply-subtract instructions.

A branch prediction algorithm is used to reduce branch penalties. Forward branches are predicted untaken and backward branches are predicted taken to optimize performance for loops. Correctly predicted branches execute in a single clock cycle.

All floating point operations except divide and square root are fully pipelined with a two-cycle latency in both single and double precision. The processor can issue an independent floating-point operation every clock cycle. Divides and square roots take 8 cycles in single precision and 15 cycles in double precision. Instruction execution does not stop for divide/square root until the result register is needed or another divide/square root is issued.

The data cache of the SPP-1000 CPUs is 1 MByte in size and is virtually indexed. It is clocked at same frequency as the CPU and is connected to the CPU/FPU through a 64-bit data path. The line size of the data cache is 32 bytes. The data cache is parity protected. A copy-back store policy is used. Data cache reads take one cycle; writes require two cycles.

The instruction cache of the SPP-1000 CPU is 1 MByte in size and is virtually indexed. It runs at the same frequency as the CPU. The instruction cache is connected to CPU/FPU through a 64-bit data path and is parity protected. If a parity error is detected, it is treated as a cache miss and the line is fetched from memory again. The instruction cache has enough bandwidth to support the continuous execution of two instructions per cycle.

The Exemplar SPP-1000 architecture supports synchronization of multiple threads of a process through the use of hardware semaphore and synchronization operators. These operators are accessible from high-level programming languages through compiler support or explicit user directives.

The Exemplar SPP-1000 architecture supports three semaphore operators: fetch and clear, fetch and increment, and fetch and decrement. These perform their respective operations without encaching the semaphore variable. If the semaphore variable was encached before the semaphore operation, the variable is flushed from the caches before the operation is performed. One additional operator, fetch, loads a semaphore variable into a register without encaching it. The semaphore operations execute atomically on memory.

In addition to the fetch semaphore operators, the load and clear semaphore instructions defined in the HP PA RISC instruction set are also supported.

Uses for the semaphore operators include nonencached binary semaphore locks, barrier synchronization, multiple reader/writer synchronization, and synchronization of queues manipulated by multiple processors.

The barrier synchronization operation is constructed with a fetch and decrement used to determine when the last thread of a process has entered the barrier synchronization, followed by a release mechanism to all threads to continue execution. The barrier synchronization is used to synchronize all threads of a process after a section of code is completed, and before the next section of code is started.

Application performance in an Exemplar SPP-1000 system is dependent on several factors. The system provides hardware to measure each of these principal factors. The measurements indicate the overall results and provide these results to development tools. These tools then enable programmers to identify possible algorithmic changes that will result in performance improvements. Some of the important factors include the following.

1. *Communication costs.* Data that are available to determine communication costs, both within and across hypernodes, include: the ratio of local-to-global memory use ratios, the memory access patterns of particular code sections, the topology of how the application is laid out across multiple hypernodes, and the communication costs between threads and/or processes.

2. *Parallel Algorithm Efficiency.* Through a combination of hardware and software, parallelization validity and efficiency can be monitored and made available. Data that are available include: trace data, synchronization statistics, granularity measurements of parallel regions, deadlock detection, and lock order enforcement.

3. *I/O Performance.* Software support for I/O performance measurement includes average and peak I/O bandwidth and distribution and the I/O load across the system.

5.8   NCUBE 3

In 1985, nCUBE introduced its first massively parallel computer. Based on a hypercube interconnect topology and the high levels of integration attained through custom VLSI technology, the nCUBE 1 system achieved good absolute performance and price/performance in scientific computing. In 1989, nCUBE introduced the nCUBE 2 family of supercomputers, a scalable series of systems capable of up to 34 GigaFLOPS and 123,000 MIPS performance. The nCUBE 2 system's high level of integration provided scalability, and cost-effectiveness for a wide range of application areas.

nCUBE offers a new family of massively parallel products—the nCUBE 3 systems (see Table 5-6). These supercomputers are designed to be multi-TeraFLOPS platforms and are compatible with previous-generation nCUBE systems. The nCUBE 3 family carries on nCUBE's philosophy of integration and scalability, offering systems that scale from low-end, entry-level products to high-end, Grand Challenge machines.

Using many techniques from the nCUBE 1 and nCUBE 2 families of systems, nCUBE 3 supercomputers deliver scalable performance through a balanced architecture. The MIMD architecture gives users programming flexibility, offering a multiapplication, multiuser environment.

By using custom VLSI technology to integrate all the components necessary for parallel computing onto a single chip, the nCUBE 3 system scales from as few as eight processors to as many as 65,536 processors in a single system architecture.

TABLE 5-6

NCUBE 3 SPECIFICATIONS	
Performance	6.5 TeraFLOPS DP IEEE format 3-million MIPS, 64-bit integer
Memory	65 TeraBytes ECC protected 30 TeraBytes-per-second aggregate memory bandwidth
Communications	24 TeraBytes per second—hypercube interconnect
I/O	1024, 3 GigaBytes-per-second channels Multiple terabytes of RAID-based storage HiPPI, Fiberchannel, FDDI support

The nCUBE 3 system delivers performance levels ranging from under one GigaFLOPS to more than 6.5 TeraFLOPS.

The nCUBE 3 system's architecture makes it suitable for large application sizes. nCUBE 3 computers are expandable in increments of eight processors. The nCUBE 3 system architecture and communications topology do not impose restrictions on scalability. Applications may be run on "subcubes" (partitions), with sizes varying from a single processor to the entire system configuration. Many subcubes of differing sizes can run simultaneously in the nCUBE 3 system, allowing resources to be scaled to match the requirements of many complex jobs.

A number of programming paradigms are supported by the nCUBE 3 system, including Single Program, Multiple Data (SPMD); cooperative processing between subcubes; and cooperative processing with client systems via network connections.

As a result of the high-density connection characteristics of the system's hypercube topology, other interconnect topologies can be emulated with virtually no performance impact. Even "shared memory" access to remote data is supported via hardware and software mechanisms for implicit message generation.

The nCUBE 3 system's Parallel Software Environment (PSE) continues nCUBE's tradition of microkernel architectures and open systems standards by combining the nCUBE nCX microkernel with the functionality of the UNIX operating system. In this way, nCUBE 3 supercomputers maintain the low operating overhead and efficient messaging mechanisms of nCX while providing complete API compatibility with standard UNIX systems—an approach that results in enhanced application portability as well as integration into distributed-computing environments.

The nCUBE 3 system supports a broad range of computational and communications libraries, easing application migration. These libraries are tuned and optimized to take advantage of the nCUBE 3 system's low latency messaging and high-performance instruction-set architecture.

nCUBE's PSE delivers a robust development environment for the nCUBE 3 system. The nCUBE 3 PSE will support standard programming languages, including FORTRAN 77, ANSI C, C++, and Ada as well as parallel languages such as HPF FORTRAN, Parallel Prolog, and Data Parallel C. Languages are implemented via nCUBE's common set of backend optimization and code-generation tools.

To ease application porting to parallel environments, third-party tools such as Forge 90, Linda, Express, and others are available on the nCUBE 3 platform.

Although shared-memory architectures (see Figure 5.8-1) are practical for systems with small numbers of processors, these architectures do not scale well as processors and memory are added. When processors are added to a shared-memory system, contention for the system's limited bus bandwidth is increased, degrading processor-memory performance. Further, when processors are added in a shared-memory system, the amount of logic (e.g., cache coherency mecha-

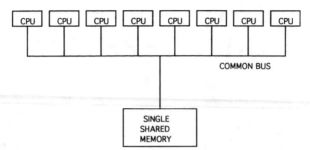

Figure 5.8-1.   Tightly coupled memory architecture. (A fixed memory bus
bandwidth limits performance.)

nisms) required to manage memory must be increased, reducing the system's performance.

Distributed-memory systems (Figures 5.8-2 and 5.8-3), however, do not suffer from these drawbacks. In a distributed memory system, each processor has
its own local memory. The potential bottlenecks associated with a processor-
memory bus and the need for elaborate cache management systems are eliminated. Adding processors adds memory; processor-memory bandwidth scales
with computational power. At the same time, the requirement for high-bandwidth
interprocessor communication is addressed by the nCUBE 3 system's interconnect topology. This topology allows processors to share data and processing.

An interconnect topology capable of scaling to large numbers of processors
without degrading communication performance or increasing latency is manda-
tory for massively parallel systems. The hypercube interconnect topology—a network of 2^n processors configured in an n-dimensional cube—meets these
requirements; nCUBE's hypercube interconnect is deadlock free, and provides
graceful degradation in the event of component failure. Hypercube interconnects
can also emulate other communication topologies, assuring that users will have
the correct interconnect topology for the task at hand (Table 5-7).

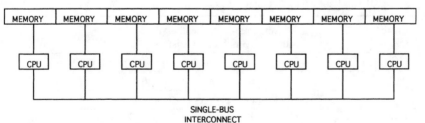

Figure 5.8-2.   Loosely coupled memory architecture. (A fixed interconnect
bandwidth limits performance.)

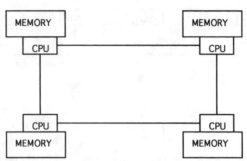

Figure 5.8-3.   In a distributed memory architecture, such as the nCUBE 3 system, each processor has its own local memory. Potential bus and interconnect bandwidth bottlenecks are eliminated.

nCUBE systems employ a custom instruction-set architecture that allows the advantages of VLSI technology to be fully realized in a massively parallel system. Off-the-shelf microprocessor architectures employ mechanisms for shared memory, vector floating-point units, and aggressive cache architectures—necessary features for the mass workstation marketplace, but superfluous for the design of massively parallel systems. Hence, off-the-shelf microprocessor architectures result in design compromises such as higher power dissipation, longer latencies for memory access, and lower levels of system integration. nCUBE processors, however, employ features such as low-latency networking, low-latency memory access, full 64-bit operation, and integrated memory control with error correction, all with very modest power dissipation (see Table 5-7).

The nCUBE 3 system uses standard, off-the-shelf DRAM, supporting as much as 1 GigaByte of local memory per processor. By using commodity memory components, nCUBE avoids complex memory architectures, resulting in systems with increased reliability and low costs.

TABLE 5-7

NCUBE 3 PROCESSOR SPECIFICATIONS	
Custom VLSI	0.6 μm, three-layer metal CMOS More than 3 million transistors
Clock Rate	50 MHz, with technology path to 66 and 100 MHz
ALU	64-bit, Virtual Memory Architecture
FPU	100 MFLOPS per processor (scalar)
Inst. Cache	16 KB
Data Cache	16 KB

The nCUBE 3 system's hypercube network of high-speed computational nodes offers balanced processor, communications, memory, and I/O performance, providing application throughput that scales linearly as the system's total number of nodes is increased. Each computational node contains a high-performance, 64-bit microprocessor, local memory, and network connections to other nodes and I/O devices.

The nCUBE 3 system contains as few as eight to as many as 65,536 computational nodes. In a maximum configuration system, the computational nodes communicate via an internal hypercube network at an aggregate data rate of 24 TeraBytes per second.

A single node's local memory scales from as little as 16 Megabytes to as much as 1 Gigabyte, and nodes of differing memory capacities may be used in the same system. However, applications need not be limited to the memory capacity of a single computational node. The nCUBE 3 system employs a virtual-memory architecture based on a 64-bit address and provides up to 256 TeraBytes of address space for each process.

To deliver superior I/O capabilities, the nCUBE 3 system features the ParaChannel I/O array, an independent hypercube network of nCUBE 3 nodes for load distribution and I/O sharing. Computational nodes in the processing array communicate via direct network links to a ParaChannel I/O array. Computational nodes may also communicate indirectly to the ParaChannel I/O array by using the system's hypercube network. ParaChannel I/O array nodes can each service the requests of up to eight computational nodes.

The nCUBE 3 communication architecture, like all binary hypercubes, doubles the size of the processor array when an additional hypercube dimension is employed. nCUBE's implementation allows for partially expanded hypercubes, enabling the processor array to be upgraded in increments of as few as eight nodes, regardless of system configuration.

nCUBE 3 computational nodes, as well as ParaChannel I/O nodes, are based on the nCUBE 3 microprocessor. This third generation of the nCUBE processor is fully compatible with previous-generation nCUBE processors. The nCUBE 3 processor integrates an entire computer system on a single chip (Figure 5.8-4).

At the core of the nCUBE 3 processor are its Arithmetic Logic Unit (ALU) and Floating Point Unit (FPU). Designed as a true 64-bit architecture, the ALU completes most integer operations in a single 20-nanosecond clock cycle. The FPU, by employing its multiply/add function, can complete two single- or double-precision floating-point operations in a single clock cycle. As in the nCUBE 2 processor, all floating-point operations are fully IEEE compatible.

Floating-point operations are implemented in a scalar fashion, which requires only a two-stage pipeline for execution, and permits codes that have relatively random data-access patterns to achieve a high percentage of the nCUBE 3 system's peak floating-point performance (see Table 5-8). Close integration with the processor's internal cache and memory interface allow the nCUBE 3 system to sustain peak performance rates with long memory resident vectors.

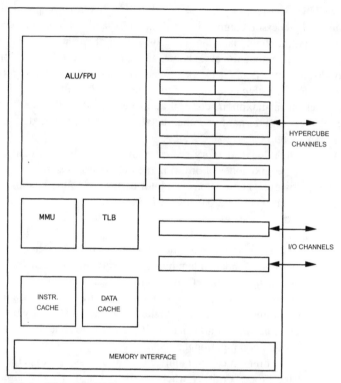

Figure 5.8-4. The nCUBE 3 processor integrates all the components for parallel computing onto a single chip.

TABLE 5-8

NCUBE 3 SYSTEM PERFORMANCE	
Specifications	**Architectural Maximum**
Processors	65,536
Main Memory (TB)	65
TeraFLOPS	6.5
MIPS	3,000,000
Memory Bandwidth (TB/s)	30
Communications Bandwidth	24
I/O	1024, 3-GigaByte channels

The nCUBE 3 processor has two on-board caches and a register file. The instruction cache and the data cache are each 16-KiloBytes deep and two-way set associative. The caches have an internal data bandwidth of 1.6 GigaBytes per second and are serviced by an independent load/store pipeline for optimal utilization. Caches are accessed by physical address to avoid unnecessary flushing of data after context switches.

The nCUBE 3 processor includes support for a demand-paged virtual-memory environment. Based on an architecturally defined 64-bit virtual address designed for future expansion, the nCUBE 3 architecture implements 48 address bits, yielding a virtual-address space of 256 TeraBytes. Page sizes may be mixed and can vary from 256 Bytes to 64 MegaBytes. A 64-entry, fully associative Translation Lookaside Buffer (TLB) is used to sustain high-performance address translation.

Although the nCUBE 3 system employs a distributed-memory design with data sharing implemented by explicit messaging, there are times when an implicit reference to another computational node's memory is useful. For instance, it is sometimes helpful to use another node's local memory as an extended data buffer instead of using secondary storage, and it is occasionally useful to achieve limited shared access to global structures. In such cases, nCUBE 3 system's virtual-memory architecture allows a page to be marked as nonresident; appropriate messages to transfer the page to the local node are then generated.

Beyond shared access to data, the nCUBE 3 system extends MIMD architecture semaphore management by using hardware-generated messages to provide atomic data operations on remote memory locations. For critical operands such as shared semaphores, this mechanism provides the fastest access possible in a distributed memory system, enhancing the speed at which applications can synchronize operations.

The nCUBE 3 supercomputer's memory architecture is designed to perfectly complement the high-speed nCUBE 3 processor. Capable of sustaining transfer rates of 500 MegaBytes per second over its 64-bit memory bus, the nCUBE 3 processor supports Synchronous DRAMS (SDRAMS) in its design. SDRAMS provide high-speed internal buffers capable of transferring data directly to the nCUBE 3 cache or arithmetic units at peak processor data rates, without memory interleaving. This approach provides flexibility in memory system configuration, without degrading performance.

The nCUBE 3 system employs a 30-bit physical address and is capable of directly addressing 1 GigaByte of local memory. All memory operations are completely ECC protected; single-bit errors are corrected and double-bit errors are detected and reported.

In total, the nCUBE 3 hypercube interconnect network has improved its average data capacity by more than a factor of 100 over nCUBE 1. Each computational node is capable of a 400 MegaBytes-per-second peak communication data rate across its data channels.

nCUBE 3 I/O employs the same features used to enhance the system's hypercube interconnect. nCUBE 3 I/O connections are designed in the same manner as the system's interconnect structure. Each nCUBE 3 processor contains two independent I/O DMA channels in addition to 16 hypercube DMA channels used for interprocessor communication. These independent I/O channels provide 20 MegaBytes per second of full duplex I/O for every computational node, and also permit alternate I/O paths for fault resilience.

The system's I/O DMA channels are directly connected to an I/O node on a ParaChannel board. By providing this direct interconnect, I/O can be scaled to a maximum of 1024 ParaChannels, each capable of more than 2.5 GigaBytes per second of I/O.

The ParaChannel I/O array is an independent, 16-node hypercube system offering peak performance of 800, 64-bit integer MIPS. Standard nCUBE 3 processors in the ParaChannel array run I/O drivers, file systems, and communications channel protocols. Individual nCUBE 3 nodes are personalized by a dual-ported memory connection to the required I/O interfaces.

The inherently parallel nature of the ParaChannel makes it suitable for direct implementation of Redundant Arrays of Inexpensive Disk (RAID) mass storage, assuring data availability and integrity. Each ParaChannel I/O node supports many SCSI disks and multiple controllers, providing 1 GigaByte per second of available disk bandwidth per channel. A single ParaChannel I/O array may be connected to more than 400 disks.

nCUBE 3 systems let users connect to high-performance networks, as well as a variety of supercomputers, mainframes, minicomputers, and workstations. The nCUBE 3 Parallel Software Environment for the nCUBE 3 system offers a variety of interfaces, protocols, and physical media, including TCP/IP and higher level protocols, as well as the High-Performance Peripheral Interface (HiPPI), the Fiber-optic Digital Data Interface (FDDI), and Ethernet.

nCUBE 3 system software is compatible with software available on nCUBE 2 systems. The nCUBE Parallel Software Environment (PSE) for the nCUBE 3 system provides a robust suite of system software, development tools, languages, and libraries that deliver the power of parallel processing in a simple package. The nCUBE PSE for the nCUBE 3 system contains the following.

- The nCX microkernel operating system
- Industry-standard UNIX functionality
- A variety of tools for writing, compiling, profiling, debugging, launching, and controlling parallel programs
- Advanced system administration utilities
- On-line documentation

Using the nCUBE PSE for nCUBE 3 systems, users can work independently, exercising autonomous control over the execution of programs. Multiple users can also run programs simultaneously.

The PSE for the nCUBE 3 is based on nCX, a highly optimized, small, and fast microkernel that executes on all compute nodes and I/O nodes in the nCUBE system. The nCX microkernel provides services for process management and memory management, as well as a high-performance messaging system, all designed to give users a productive work environment. Other operating system services and facilities, such as I/O and networking, may be distributed across any number of compute or I/O nodes in the system.

The Parallel Software Environment for nCUBE 3 systems provides a variety of programming languages, including FORTRAN 77 with VAX extensions, FORTRAN 90, ANSI C, C++, and Ada, as well as parallel languages such as HPF FORTRAN, Parallel Prolog, and Data Parallel C. All languages are supported with standard compile-time and run-time constructs. nCUBE libraries extend these languages to provide communication and data decomposition routines.

5.9   SILICON GRAPHICS— POWER CHALLENGEARRAY

POWER CHALLENGEarray is a distributed parallel processing system that scales up to 144 MIPS R8000 or up to 288 MIPS R1000 microprocessors to serve as a powerful distributed throughput engine in production environments, and to solve grand challenge-class problems in research and production environments. POWER CHALLENGEarray consists of up to eight POWER CHALLENGE or POWER Onyz (*POWERnode*) Supercomputing systems connected by a high-performance HiPPI interconnect. Using the POWERnode as a building block, POWER CHALLENGEarray exploits the ultra-high-performance *POWERpath-2*™ interconnect to form a modular, scaleable system. Using this unique modular approach, POWER CHALLENGEarray creates a highly scalable system, *providing more than 109 GFLOPS of peak performance, up to 128 GB of main memory, more than 4GB/sec of sustained disk transfer capacity, and more than 28TB of disk space.*

POWER CHALLENGEarray therefore offers a two-level communication hierarchy, where CPUs within a POWERnode communicate via a fast shared-bus interconnect, and CPUs across POWERnodes communicate via a high-bandwidth HiPPI interconnect. Each POWERnode comes with a suite of parallel programming software tools already on POWER CHALLENGE (e,g., the MIPSpro compiler, CHALLENGEcomplib, and ProDev/Workshop Application Development Tools). Additionally, POWER CHALLENGEarray offers other software tools to aid in developing distributed parallel programs, as well as to manage the parallel computer from a single point of control.

In contrast to traditional distributed-memory private-address-space architectures, the hierarchical POWER CHALLENGEarray approach offers several advantages.

- Improves computation-to-communication ratio of parallel applications
- Reduces physical memory requirements of applications
- Provides better load balancing for irregular problems
- Improves latency tolerance for coarse-grained, message-level parallelism
- Offers a flexible parallel programming environment, rich in several types of parallel programming paradigms
- Provides a gradual learning path involving shared, distributed, and distributed-shared-memory (hybrid) parallel programming for programmers relatively new to parallel programming

The POWER CHALLENGEarray approach is unique among distributed computing models because each individual system comprising the array is itself a parallel supercomputer. POWER CHALLENGEarray combines the efficiency, flexibility, and ease of programmability features of shared-memory multiprocessing with the upward scalability of message-passing architectures. This leads to a better computation-to-communication ratio because messages are sent between systems for larger blocks of parallel processing.

Each POWERnode of POWER CHALLENGEarray is a Silicon Graphics POWER CHALLENGE shared-memory multiprocessor system, supporting up to 18 MIPS R8000 or 36 MIPS R10000 supercomputing microprocessors and can provide over 13.5 GFLOPS of peak performance. An interleaved memory subsystem accepts up to 16 GB of main memory, and a high-speed I/O subsystem scales up to four 320 MB/second I/O channels. Each POWERnode system also supports a wide range of connectivity options, including HiPPI, FDDI, and Ethernet™.

If the number of tasks required in a parallel program is less than or equal to the number of CPUs on a POWERnode, the shared-memory, single-address-space programming model can be used. This provides an intuitive and efficient way of parallel programming to the user and avoids the overhead of message-passing between processes. Since there are a large number of such jobs in a typical parallel environment, POWER CHALLENGEarray can be used as an efficient throughput engine for such a job mix. Additionally, users can access the entire set of array tools and resources for POWER CHALLENGEarray.

Applications with large computational, memory, and I/O requirements that cannot be accommodated on individual workstation-class machines (that form the basic building blocks of traditional distributed-memory parallel machines), and that can exploit a moderate level of parallelism are particularly well-suited for POWER CHALLENGEarray.

Applications having scalability or memory requirements beyond the capabilities of a single POWERnode can be restructured using hierarchical programming techniques to span multiple POWERnodes. Under this model, the tasks within a POWERnode still communicate via shared memory and tasks between POWERnodes communicate via message-passing. The message-passing overhead can be optimized, compared to the computation for each message sent, since parallel tasks within a POWERnode can use global shared memory to communicate shared data.

For many applications, domain decomposition results in maximum data locality and data reuse, resulting in reduced intertask communication.

The programmer has the flexibility to use a single message-passing library for communicating within and across POWERnodes. This is because message-passing libraries use low-latency, high-bandwidth shared-memory techniques to communicate data within POWERnodes and high-bandwidth networking protocols to communicate across POWERnodes. Thus, POWER CHALLENGEarray offers the great advantage of low communication-to-computation ratios for many large-scale problems while involving no more work than traditional distributed-memory, private address-space systems.

Most real applications have some amount (in the order of 5% or more) of nonparallelizable code. Using more than a moderate number of processors (10–20) for them does not yield additional performance benefits. For such an application, running on a single POWERnode with a moderate number of processors may be sufficient to realize any potential benefit from parallelization. Hence, a typical parallel environment will consist of some large-scale parallel applications mixed with a large number of modestly parallel applications.

For a workload consisting of moderately parallel applications, POWER CHALLENGEarray systems deliver very high throughput. This is because most of these applications can be parallelized using shared-memory techniques and run within a POWERnode. Combining this with Silicon Graphics parallelizing tools, the IRIX operating systems, and the batch processing tools available, POWER CHALLENGEarray serves as a distributed parallel throughput environment.

POWER CHALLENGEarray supports fine-grained, medium-grained, and coarse-grained parallelism. It also supports both shared-memory and message-passing programming models, and several hybrid combinations of those two models. Shared-memory programming typically involves the usage of the parallelizing FORTRAN and C compilers, whereas message-passing programming typically involves the usage of popular message-passing communication libraries such as the Message-Passing Interface (MPI) Standard and Parallel Virtual Machine (PVM). Applications may alternatively use High Performance FORTRAN (HPF) as their parallel library.

Most applications with fine and medium-grained parallelism can be efficiently parallelized using easy-to-use shared-memory techniques available on each POWERnode system. This system enables a wide range of scientific, engineering, and commercial applications to take advantage of shared-memory parallelism by using parallelizing FORTRAN 77, FORTRAN 90, and C compilers.

Applications with medium- and coarse-grained message-level parallelism can use the efficient communication characteristics of shared-memory within each POWERnode and resort to conventional message-passing between POWERnode systems. Also, many message-passing applications previously running on workstation clusters or pure message-passing architectures can run efficiently on a single POWERnode. Common message-passing libraries such as MPI and PVM on POWER CHALLENGEarray exploit the fast shared-memory communication path between tasks on a POWERnode.

For applications with large scalability and memory requirements, the algorithm can be restructured to use the hierarchical programming model to combine the benefits of shared memory with the upward scalability of message-passing techniques. POWER CHALLENGEarray therefore supports several parallel programming models, including the following.

- Shared-memory with n processes inside a POWERnode
- Message-passing with n processes inside a POWERnode
- Hybrid model with n processes inside a POWERnode, using a combination of shared-memory and message-passing
- Message-passing with n processes over p POWERnodes
- Hybrid model with n processes over p POWERnodes, using a combination of shared-memory within a POWERnode system and message-passing between POWERnodes

POWER CHALLENGEarray offers a variety of software tools, enabling it as a distributed parallel processing system and as a distributed parallel-throughput processing engine. Each POWERnode is loaded with the original tools suite from the POWER CHALLENGE platform, including the following.

- *XFS™* high-performance, journaled filesystem
- *NFS™* Version 3 network filesystem
- *MIPSpro Power FORTRAN 77, MIPSpro Power FORTRAN 90* and *MIPSpro Power C* compilers support automatic/user-directed parallelization of FORTRAN 77, FORTRAN 90, and C applications for shared memory multiprocessing

- *CHALLENGEcomplib*, a comprehensive collection of scientific and math subroutine libraries that provide support for mathematical and numerical algorithms used in scientific computing
- *ProDev/Workshop*, a suite of software development tools that includes a parallel program development tool, a parallel debugger, a parallel program profiler, and performance-tuning tools

The system provides additional tools to aid in distributed program development and distributed system management, including the following.

- HiPPI high-performance networking software
- Array services, array management software
- IRISconsole centralized server administration software
- Message-Passing Interface (MPI) communication software
- Parallel Virtual Memory Machine (PVM) communication software
- Array diagnostics

A system consists of up to eight POWERnodes connected by high-performance interconnection technology in ways that can be customized to suit communication requirements of varied problems. HiPPI is the recommended interconnection technology. The high-bandwidth characteristics of multiple HiPPI channels (100 MB/sec per channel) make HiPPI apt for the relaying of large amounts of data between POWERnodes. Also, HiPPI uses ANSI standard-conforming protocol networking technology, assuring interoperability with other HiPPI equipment.

A switch-based interconnection topology is recommended to connect POWERnodes. A switch-based system can dynamically shift between different topologies (1D ring, 2D mesh, 3D Torus) to conform to the application's communication requirements.

POWER CHALLENGEarray has a highly flexible and scalable interconnection architecture that need not restricted to a single topology, technology, or predefined interconnect bandwidth. The interconnection topology can scale incrementally with additional POWERnodes, and is easily replaced when new interconnect technology, such as ATM, becomes available. All POWERnodes run Silicon Graphics IRIX 6.1 enhanced 64-bit UNIX operating system which includes a multithreaded kernel, the XFS high-performance journaled filesystem, and NFS version 3 networking software (Table 5-9).

Each POWERnode is a shared-memory supercomputer, consisting of multiple R8000 CPU boards, interleaved memory cards, and POWER Channel™ 2 I/O boards, together with a wide variety of I/O controllers and peripheral devices.

TABLE 5-9

POWER CHALLENGEARRAY MAXIMUM SYSTEM CONFIGURATIONS	
Component	**Description**
Peak Performance	Over 109 GFLOPS
POWERnodes	8
Processor	288 MIPS R10,000 or 144 MIPS R8000
Main Memory	128 GB
Bisection BW	1.6 GB/sec
Disk I/O BW	4 GB/sec
RAID Storage Capacity	139.2 Terabytes

These boards are interconnected by the POWERpath-2 bus, which provides high-bandwidth, low latency, cache-coherent communication between processors, memory, and I/O graphics subsystems (Table 5-10).

The High-Performance Parallel Interface (HiPPI) is the recommended inter-connection technology for POWER CHALLENGEarray. Each POWERnode is required to have one or more bidirectional HiPPI interfaces. HiPPI is the industry standard for high-bandwidth networking today in both system-to-system and system-to-peripheral environments. Standardized by the American National Standards Institute (ANSI), HiPPI is widely adopted by research, higher education,

TABLE 5-10

POWERNODE MAXIMUM SYSTEM CONFIGURATIONS	
Component	**Description**
Processors	18 MIPS R8000 CPUs 36 MIPS R10,000 CPUs
Peak Performance	Over 13.5 GFLOPS
Main Memory	16 gigabytes, 8-way interleaving
I/O bus	Six POWER Channel-2 buses, each providing up to 320 MB/sec I/O bandwidth
SCSI channels	40 fast-wide independent SCSI-2 channels
Disk	17.4 TB disk (RAID) or 5.6 TB non-RAID disks
Connectivity	Six HiPPI channels, eight Ethernet channels
VME slots	Five VME64 expansion buses provided 25 VME64 slots

and engineering organizations worldwide. It is a simplex point-to-point interface for transferring data at peak data rates of 100 or 200MB/sec over distances of up to 25 meters. IRIS® HiPPI, which provides HiPPI connectivity for Silicon Graphics machines, supports the 100 MB/sec option.

The HiPPI physical layer specifies 50-pair, twisted-pair cables for distances up to 25 meters, with the 100MB/sec option using one cable and the 200 MB/sec option using two cables. HiPPI signal lines are unidirectional to accommodate fiber-optic implementations and crossbar switches. Control and data signals are timed with respect to the continuous 25 MHz clock signal. IRIS HiPPI contains two 32-bit parallel channels clocked at 25 MHz. In addition to the Framing Protocol (HiPPI-FP), which is used by the Silicon Graphics implementation of the MPI library, TCP/IP can be layered over HiPPI, providing a fast communication fabric for TCP/IP applications while retaining naming, reliability, and internetworking flexibility.

Chapter 6

Alternative Architectures

An eclectic set of five out-of-the-mainstream parallel supercomputing architectures are presented in this chapter.

1. The Hecht-Nielsen Corporation's SNAP is a neural net computer.

2. ATCURE, ERIM's Advanced Target CUeing and Recognition Engine combines four processor subsystems, each having a distinct architecture (image processing pipeline, numeric processor, symbol processor, and input/output processor).

3. A systolic architecture termed WARP developed at Carnegie-Mellon University.

4. Tera's multithreaded architecture.

5. The Image Understanding Architecture, a heterogeneous pyramid architecture developed at the University of Massachusetts.

One or more of these machines may be a harbinger of directions to come in the future because architectural diversity is expected to increase, not with large numbers of special purpose computers, but with a relatively small set of architectures designed for classes of domain-specific algorithms.

6.1 HNC SNAP

Large neural network problems typically involve an enormous number of matrix-vector operations. In order to increase the processing speed to a level where solutions are possible, many processors can be organized to simultaneously perform matrix-vector operations. HNC has developed a Single Instruction Multiple Data

(SIMD) Neurocomputer Array Processor (SNAP) that uses IEEE 32-bit floating point processing elements. The prototype SNAP is composed of a SIMD array of between 16 and 64 parallel processing elements. Each processing element consists of a 32-bit floating point multiplier, a 32-bit floating point arithmetic logic unit, and a 32-bit integer arithmetic logic unit. Four processing elements are integrated onto a single VLSI gate array. Each chip in the SIMD array is capable of simultaneously performing 160 million floating point operations per second. This gives an array of eight chips a peak processing performance of 1280 megaflops. By using VLSI technology to produce four 32-bit processing elements per chip and off-the-shelf bit-slice components to assemble the SIMD controller, this compact array processor provides a new level of performance to cost and is suitable for many applications in addition to neurocomputing.

HNC's SIMD Neurocomputer Array Processor (SNAP) architecture is most efficient for the matrix-vector operations that characterize neural networks as well as many other applications. HNC's goal has been to provide a coprocessor that is fast enough for real-time neural network applications, while providing the programmability and flexibility required for a variety of other applications in a research and development environment. The SNAP chip provides its processing power in a system with a flexible programming environment. Data and weights are treated as 32-bit floating point numbers in standard IEEE format. Since the data are represented as floating point numbers, there are no algorithmic problems with precision and dynamic range.

A consideration that heavily influenced the SNAP decision was the requirement to provide the flexibility necessary for the many neural network algorithms in use and those yet to be invented. In addition, many applications require extensive conventional preprocessing such as Fourier transforms, convolution, scaling, and so on. For real time applications, the complete program must be treated, not only the neural network portion.

The result of the analysis of the neural network algorithms and the applications in which these algorithms would be used pointed very strongly toward the development of a general purpose parallel architecture. HNC reached the same conclusion that many others had [1-9], that a ring systolic array of SIMD processors was a very powerful architecture for neural networks. After adopting that architecture, HNC diverged from other designs and implementations and concentrated on making the processing element as general purpose, as powerful, and as flexible as possible. It is relatively easy to design a machine that computes the backpropagation algorithm fast and efficiently. It is more difficult and much more important to design and build a machine that will be flexible enough to solve complete problems.

The SIMD Neurocomputer Array Processor system [10] is composed of three parts: the SNAP array of VLSI chips, the SNAP Controller, and a host interface processor. In addition, since the SNAP operates as a coprocessor, a host computer such as a Sun SPARCstation is also required. The configuration for a system with 32 processing elements is illustrated in Figure 6.1-1. HNC designed and developed the new SNAP VLSI chip to exploit the inherent massive parallelism in neural networks. A system built from these chips is constructed on 6U VME boards and consists of a minimum of two boards: a general purpose coprocessor board with an Intel i860 processor and 16 MBytes of memory, plus a Linear Array board with four SNAP chips. A SNAP Controller daughterboard sits on top of the Linear Array board. The system is extensible such that four Linear Array boards can be connected together and controlled by a single SNAP Controller daughterboard. A system with two Linear Array boards will operate at 1280 megaflops peak performance. The SNAP system block diagram with four Linear Array boards as it would be used in the Sun SPARCStation is shown in Figure 6.1-1.

Each SNAP chip contains four processing elements with each one connected to its neighbor in a linear array. A system with two Linear Array boards will have eight SNAP chips and thus 32 processing elements. All the processing elements are controlled by a single instruction from the external controller. Each processing element in the chip can perform both floating point and integer arithmetic,

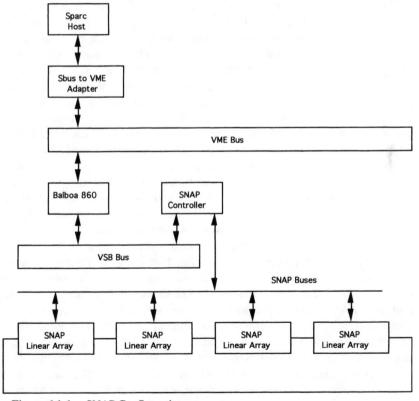

Figure 6.1-1.   SNAP Configuration.

and is connected to two external memory banks. One of the memory banks is local memory and the other bank is global memory.

Each processing element has a 32-bit path to its local memory, a 32-bit path to the next element in the array, a 32-bit path from the previous element in the array, and a 32-bit path to the global memory bus. The global memory bus is connected to all the processing elements in the array and to global memory. The bus is a broadcast bus such that data in the global memory can be broadcast to all processing elements at the same time. Alternatively, data in any single processing element can be broadcast to all other processing elements.

Three separate parallel arithmetic units exist in each processing element: a 32-bit floating point multiplier, a 32-bit floating point arithmetic logic unit, and a 32-bit integer arithmetic logic unit. The floating point units support the ANSI/IEEE Standard P754-1985. There are no internal pipelines so that all instructions execute in a single cycle. This absence of internal pipelines simplifies the programming task. Each of the arithmetic units has a separate instruction input and operates independently. However, the floating point ALU and the integer ALU cannot execute in the same cycle. Each processing element contains a total of seven data registers, three instruction registers, and one status register.

The local memory is composed of a block of 32-bit static RAM connected to each processing element by a bidirectional data bus. An output register and an input latch are provided inside each processing element for maximum flexibility. The processing element drives or reads the data bus on direction from the instruction bus. Each processing element provides a local memory enable signal to the local memory indicating whether it is active for this transaction. The local memory may be conditionally enabled on the results of local processing. The address bus of the local memory is driven by the SNAP bit-slice computer and can be made an arbitrary size, up to 32 bits. HNC has implemented an address bus of 17 bits on the prototype 6U cards, giving each processing element a local memory of up to 512 Kbytes of static RAM. The size of the address bus is entirely determined by the external SNAP Controller. The SNAP chip and system architecture does not limit the size of the local memory. A system with larger boards or denser memory chips can be built as memory technology improves or as demand requires.

The global memory is connected to each processing element by a bidirectional data bus. Each processing element interface has an output register and an input latch for maximum flexibility. The global memory consists of a 32-Kbyte block of 32-bit dual port static RAM and a 512-Kbyte block of conventional single port static RAM. The dual port memory serves both as a means of storing global data and as an effective way of communicating to the Balboa 860 board over the VSB bus. The dual port memory has one address bus driven by the SNAP Controller and the other address bus driven by the VSB interface. The address of the 512 Kbytes of additional global RAM is driven only by the SNAP Controller.

Each processing element in the array is connected to its neighbor on the left and its neighbor on the right by bidirectional 32-bit data buses that are used to shift data between processing elements. The output of the last processing element in the array is connected to the input of the first processing element making a ring out of the linear array. A data shift operation can take place in parallel with arithmetic operations in the same cycle. This capability gives the SNAP architecture both a SIMD quality and a systolic quality. In addition to nearest neighbor connectivity, each processing element can communicate through the global bus to all other processing elements or a selected set of processing elements. Only one processing element can transmit on the global bus during a given cycle. All the other processing elements listen. However, each processing element has an activation control line that comes from the SNAP Controller that tells it whether to halt or whether to execute the present instruction. This set of lines can be used to designate a selected set of processing elements that will operate on the data on the broadcast bus. The particular case where the set of operating processing elements has only one member gives the SNAP architecture the ability to communicate data between any two processing elements.

An example where the point-to-point connectivity is particularly useful occurs in the neural network backpropagation algorithm. In the learning mode, each local processing element computes a local partial sum of error terms for the neurons that it has in its local memory. In order to obtain the complete error term, the partial sums must be collected and accumulated from all the processing elements. In a SNAP system with 32 processing elements, this can be done by designating the first processing element to be the accumulator and setting its instruction to accumulate. All the other processing elements shift their data 31 times until it has all passed through the first processing element. The point-to-point connectivity gives us an attractive alternative. Every eighth processing element can be set to accumulate. Thus there will be four processing elements that will accumulate. After the data are shifted seven times there will be four partial accumulations. Only three additional cycles are then required to send the partial accumulations to the first processing element and accumulate the final result there. Thus a total of 10 cycles was used instead of 31 to perform the accumulation operation.

The SNAP array is microcode controlled by the SNAP Controller. Physically, the Controller resides on a daughter board that rides piggyback on the first Linear Array board in the system. The Controller is responsible for generating the instruction stream sent to the processing elements. It is also responsible for generating the addresses and control signals for both local and global memory. To accomplish this task with minimum overhead, a powerful 16-bit microprogrammable microprocessor has been developed using an IDT49C410 microsequencer and two IDT49C402 microprocessors that are used to create two Registered Arithmetic Logic Units (RALUs).

The Controller data flow is organized around a main 16-bit data bus that connects all of the Controller's processing elements together with registers interfacing to the Linear Array, Local Memory, and Global Memory.

Programming the Controller is done using a macroassembly language developed by HNC. This assembler incorporates powerful high order constructs such as IF-THEN-ELSE statements usually only found in high order languages. By using this language to develop application programs for the SNAP, a programmer will be able to code in a more familiar, friendly environment while running the efficiency of traditional microcoding. A package of standard routines callable from the Balboa 860 in "C" has been developed to even further ease programming. These routines include matrix and vector operations in addition to the neurocomputing functions such as backpropogation.

The SNAP board set prototype is configured as a slave processor to HNC's Balboa board in a Sun SPARCStation host. The Balboa board, which is based on the Intel i860 microprocessor, has 16 MBytes of memory and is provided with a complete set of software tools including a C compiler, assembler, linker, librarian, simulator/debugger, executive, and loader.

In a neural network with Kohonen learning, only the neuron with the smallest Euclidean distance between its weight vector and the input vector is permitted to adjust its weight vector. This algorithm is executed on the SNAP as follows.

1. At the end of the dot product calculation, each processing element places the result into the systolic data register (SD) and into a general purpose register (R1A).

2. Shift the contents of the SD to the right.

3. Subtract the contents of the SD register from the contents of R1A.

4. If the result of the subtraction is less than 0.0, then halt execution of the arithmetic units.

5. Repeat steps 2, 3, and 4 until finished.

This procedure is executed for all the neurons in the Kohonen layer. All three steps (2, 3, and 4) are executed in parallel in the same clock cycle. A very simple case would spread a layer of 32 neurons uniformly across a SNAP system with 32 processing elements. After 31 executions of the steps 2, 3, and 4 in the preceding algorithm, only the winning neuron (or neurons if two Euclidean distances are equal) is left as an active processing element. It then is allowed to "learn" and adjust its weight values.

The first component of the software required for the use of SNAP in general applications is a general-purpose Scientific Algorithm Library (SAL). This will include three types of function. The first type will be array data manipulation functions that can serve as primitives for general programming and other SAL functions. These will consist mostly of array and matrix manipulation and elementary arithmetic functions. The second type will be the typical scientific computing functions such as transcendental and special function, FFT, convolution, eigenvector/eigenvalue, and singular value decomposition. The third type will consist of more sophisticated scientific computing functions such as differential equation solutions. The SAL will provide easy programmer access to the parallelism of the SNAP simply through a set of function calls.

One of the major benefits of supercomputers is their ability to solve problems that could not be attempted in the past because of the long execution times on more conventional machines. Now that it is possible to build these large machines, the major barrier to their widespread use is their cost. The SNAP has a tremendous advantage in this area.

In general, the task of using SIMD machines effectively depends on the efficient routing and mapping of data. Those problems that have a high degree of data parallelism can be programmed efficiently on SIMD machines. The signature of such a problem is the ability to formulate the solution in terms of identical operations on large amounts of data. Efficient matrix and vector operations are a common feature in SIMD solutions to a very large class of applications problems. Both numerical and discrete problems are typically encountered.

SIMD machines are used for applications from such diverse fields as physics, image processing, orbit analysis, operations research, and data base text search. Physics applications include fluid dynamics, electromagnetic wave scattering, thermal diffusion, seismology, and the solution of partial differential equations such as the wave equation. Image processing problems involving pattern recognition and convolution also have been programmed to take advantage of the data parallelism on SIMD machines. Operations research problems arise in a wide range of engineering, management, statistical, economic, and other applications. Such problems can involve optimization with thousands of variables and constraints. Solutions to these problems have been implemented on most SIMD machines. Data base text search also has the structure of identical operations on a large volume of data. Neural networks are a relatively recent addition to the class of applications efficiently addressed by SIMD architectures.

References

1. Kung, S.Y. and Hwang, J.N., *Parallel Architectures for Artificial Neural Nets,* IEEE International Conference on Neural Networks, San Diego CA, Vol. 11, pp. 165–172, July, 1988.

2. Pomerleau, D.A., Gusciora, G.L., Touretzky, D.S., and Kung, H.T., *Neural Network Simulation at Warp Speed: How We Got 17 Million Connections Per Second,* IEEE International Conference on Neural Networks, San Diego CA, Vol 11, pp. 143–150, July, 1988.

3. Hammerstrom, D., *A VLSI Architecture for High-Performance, Low-Cost, On-Chip Learning,* International Joint Conference on Neural Networks, San Diego, CA, Vol.11, pp. 537–544, 1990.

4. Elias, J.G., Fisher, M.D., and Monemi, C.M., *A Multiprocessor Machine for Large-Scale Neural Network Simulation,* International Joint Conference on Neural Networks. Seattle WA, Vol. 1, pp. 469–474, 1991.

5. Hirose, Y., Anbutsu, H., Yamashita, K., and Goto, G., *A VLSI Architecture for a Dedicated Back-Propagation Simulator,* International Joint Conference on Neural Networks, Seattle WA, poster presentation, 1991.

6. Means, R.W. and Lisenbee, L., *Extensible Linear Floating Point SIMD Neurocomputer Array Processor,* International Joint Conference on Neural Networks, Seattle WA, Vol. 1, pp. 587–592, 1991.

7. Ramacher, U., Beichter, J., and Bruls, N., *Architecture of a General Purpose Neural Signal Processor,* International Joint Conference on Neural Networks, Seattle WA, Vol. 1, pp. 443–446, 1991.

8. Ramacher, U., Beichter, J., Raab, W., Anlauf, J., Bruls, N., Hachmann, U., and Wesseling, M., *Design of a lst Generation Neurocomputer, VLSI Design of Neural Networks,* Ramacher, U., and Ruckert, U., (eds). Kluwer Academic Publishers, Boston, 1991, pp. 271–310.

9. Vlontzos, J.A. and Kung, S.Y., *Digital Neural Network Architecture and Implementation, VLSI Design of Neural Networks,* Ramacher, U., and Ruckert, U., (eds). Kluwer Academic Publishers, Boston, 1991, pp. 205–227.

10. Means, Robert and Lisenbee, Layne, *Floating Point SIMD Neurocomputer Array Processor*, HNC Inc., San Diego, CA.

6.2   ATCURE

There are various applications of image processing, each of which engenders different sets of requirements. Real-time interpretation of images collected from sensors on mobile platforms while maneuvering in an unconstrained or hostile environment remains one of the most challenging of these applications. This is true even though there have been great improvements over the last two decades in sensors, processor technologies, and algorithms. In this application domain the requirements for small size and weight are superimposed on the computational requirements of computer vision.

Although there are some applications within this domain that can use general purpose microprocessors, most require some type of specialized parallel image processor for the foreseeable future. Parallel processors are needed because the size of the data sets are large, they must be processed in real-time, and the algorithms for processing them are complex. Specialized parallel processors are needed because a system with sufficient computational power built only from general purpose processors cannot satisfy the weight, power, and size constraints using available fabrication technologies. To guide the development of the next generation of processors for this application domain, careful analysis of the computational requirements is required.

An embedded image processing application program can be divided into four major areas, image manipulation, numeric data processing, symbolic processing, and external communications. These categories are distinguished by the following.

1. The way data are usually organized in memory

2. The operations that are performed on the data

3. The way data values are represented

4. The control structures commonly used

Image manipulation uses images as individual variables. Images are most often (but not always) represented using integer (fixed point) values for each pixel. Vectors of integers can be used for multispectral and complex-data images. They are stored in memory so that adjacent pixel elements are in predictable locations relative to each other. Common image sizes range from 128×128 to 2 k $\times$ 8 k pixels. Typical operations involve nested loops to create new pixel values based on arithmetic operations between pixels with the same indexes in two images or between neighboring pixels in one image. Rarely are random access, shuffling, or sorting performed. Relatively infrequent exceptions would include geometric warping and the Hough Transform.

Numeric data processing uses vectors and matrices as individual variables. They are frequently represented as arrays of floating point numbers. As with images, the storage location corresponds to the index(es) of each element of the matrix or vector. In practical applications the dimensions of the matrixes are from 1×1 (scalars), to 64×64 (matrices), or 1024×1 (vectors). The number of elements in any one matrix or vector is rarely much more than 1000. Typical operations include vector-matrix products and similar multiply-accumulate operations.

Symbolic processing involves manipulation of data structures containing categorical and numerical information. Individual variables often include lists of information about each object or situation. Symbolic data are often stored in linked lists. Tree structures, graph structures, and semantic nets are often used. The categorical information within the structures is most often represented as point numbers. For symbolic data, storage locations do not correspond to locations in the image, thus the data are no longer iconic. The goal of this kind of data processing is to develop and interpret a compact description of the scene. Typical operations involve schema instantiation and other methods for matching an unknown graph to a set of models or templates [1, 2]. These decision-intensive search functions often dominate the operations on this type of data.

External communications involve interfaces to sensors, display systems, and other processing systems. Their operations are driven by the need for rapid response to external events and are often interrupt-response driven. Data structures are messages that are transferred to or from an external device. The messages can contain any of the other data types.

Achieving real-time image interpretation on mobile platforms requires a processing system that is small and light that can run application programs that include all four of these categories of processing. For most problems, a single processor is inadequate. Furthermore, for many problems in this application domain, current technology does not even permit the manufacture of a small homogeneous multiprocessor system that is capable of executing the required operations in real time. Too many processors would be required, resulting in a large volume, weight, and power requirement. It becomes necessary, particularly for image manipulation, to provide specialized processors that can exploit some of the characteristics of each processing category. Furthermore, different specialization is required for each category because of the differences in the data structures and types of operations. These distinct processors can then operate as a team of specialists to implement applications programs.

In this section the subsystem for image manipulation that is part of a heterogeneous system that includes separate processing subsystems for each of the major processing categories is described. This system is termed ATCURE, developed by ERÍM. This heterogeneous processing architecture provides the generality necessary for a wide variety of applications by integrating four distinct subsystems; one each specialized for processing of external communications, symbolic data, numeric data, and images. Each subsystem is optimized for the data types, control structures, and operation mix typical of that domain, with each subsystem configurable to include parallel processors connected via local buses. A general purpose bus and special image buses provide communication paths between the subsystems. This use of separate buses within and between the subsystems allows both feedback and feed-forward interactions.

Figure 6.2-1 depicts how the subsystems relate in a typical application program. External communications are supported by the Input/Output Processing Subsystem (IOP), which has general purpose processors to control specialized interface circuits for sensors, displays, and communications with other systems. The Numeric Processing Subsystem (NPS) processes floating point vectors and

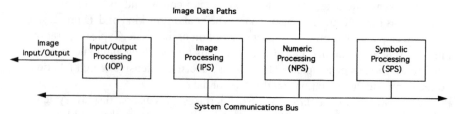

Figure 6.2-1.   ATCURE System Architecture.

matrices using an MIMD configuration of processing elements developed for digital signal processing. The Symbolic Processing Subsystem (SPS) includes general purpose processing elements, each with a large amount of local memory for program and data storage, and connected within the subsystem via a general purpose bus structure. The Image Processing Subsystem (IPS) consists of a pipeline of processing elements operating as a systolic processor connected via specialized image data paths.

One goal of the program is to combine some of the best features of previously developed systems with new solutions to reduce image manipulation bottlenecks. Some of the characteristics that have been improved follow.

- Image data access overhead
- Instruction fetch overhead
- Temporary storage requirements
- Data dependent operations implementation
- Standard reference image generation
- Labeling and other data propagation operations implementation
- Iconic to symbolic/numeric data extraction

To accomplish these improvements an analysis was performed by implementing computer simulations of the ATCURE architecture using the SES/Workbench [3] simulation environment. These simulations were performed to determine how to best incorporate features that exploit various characteristics of the data structures, mix of operations, and control structures common to this type of processing. The most significant characteristics that can be exploited are the following.

1. Images are formed from large numbers of pixels.

2. Similar operations are executed on all of the pixels selected from an image.

3. Image data can be stored so that they are usually accessed in some predictable order.

4. Neighborhood operations are about as common as multi-image operations.

5. There are often long sequences of operations that do not depend on global (image wide) analysis of the data.

Various architectural approaches have been developed to exploit these characteristics. Two common approaches are single instruction stream multiple data stream (SIMD) meshed arrays [4] and multiple instruction stream single data stream (MISD) systolic processors [5]. Each of these approaches tries to exploit the fact that similar instructions are executed on all the pixels selected from an image. In the SIMD meshed array each processing element is assigned a subset of the image and the instruction stream is broadcast to all the processing elements simultaneously. With this approach, as the images get larger, program execution time can be kept small by increasing the number of processing elements. However, once the pixels are assigned one to a processing element, execution time is proportional to the algorithm complexity. In the MISD systolic processors each instruction is executed by a different processing element and the data are pipelined between them. With this approach, the execution time is dependent on the size of the image. However, as the algorithm becomes more complex it is possible to add processing elements to keep the execution time small. In this case, once the instructions are each assigned to individual processing elements, execution time becomes proportionally to image size.

For the ATCURE Image Processing Subsystem a generalization of the MISD systolic processor architecture was developed. This approach to scalability is compatible with the raster scan nature of most sensors, data enter and leave the processor one row or column at a time. It also allows the same set of processing elements to be fully utilized for early processing of entire images and during processing of subsets of the image once regions of interest are identified for further processing. Often during early processing fairly simple algorithms are executed on the entire image, whereas processing of regions of interest usually requires more complex algorithms. The number of processing elements can be chosen to balance the time to execute simple algorithms on large images with the time to execute more complex algorithms on many smaller image regions.

The Image Processing Subsystem for ATCURE is specifically designed to be part of a heterogeneous computer system. The key feature in this subsystem is that images are the fundamental unit of data. Other types of data are expected to be processed in different subsystems. With this approach

1. A single memory access retrieves or stores an entire image, or part of an image, independent of the image size.

2. The basic instructions include neighborhood operations on single images and a complete complement of operations on pairs of images.

3. Each instruction performed by the functional units operates on an entire image or part of an image.

4. The registers within the processor store image data.

For added parallelism, the processor includes many functional units and is deeply pipelined so that many of these image-oriented instructions execute simultaneously. Finally, the processor is organized so that data flow between the registers and the functional units with minimum contention for the routing resources.

The Image Processing Subsystem includes an Image Processor Controller, Image Memories, Pipeline Processing Elements, and Feature Extractors, as shown in Figure 6.2-2. The programming and execution model implemented in the IPS is a generalization of features found in the Cyto-HSS [6]. In the IPS the processing elements operate as systolic processors connected via specialized image data paths. This provides an ultradeep synchronous pipeline of functional units that eliminates many of the bottlenecks found in MIMD and SIMD architectures. An ATCURE system with 20 functional units in the pipeline is capable of performing 8×10^9 mathematical operations per second on images with up to 2000×2000 16-bit pixels. Once address calculations, conditional branches, and data fetch instructions are included this is equivalent to over 4×10^{10} RISC instructions per second.

Individual memory, neighborhood, and multi-image instructions that can be executed by the IPS are shown in Table 6-1. In the programming model implemented by the IPS a single memory instruction causes a complete image or subimage to be read from or written to intelligent Image Memories. In typical applications the order of data storage and retrieval is known *a priori*. Furthermore, there is usually an iterative pattern in the sequence. The intelligent Image Memories include programmable sequencers that can exploit this *a priori* information about the pattern to read and write the selected pixels in a sequence compatible with the instructions executed in the functional units. The memories also include an indirect addressing feature whereby one part of the memory is used to designate the access sequence for a second part. This is used if a calculated sequence or complicated access pattern is required. It also facilitates implementation of very large look-up tables. Furthermore, since operations between images are common,

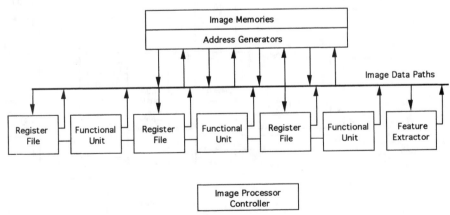

Figure 6.2-2.   The Image Processing Subsystem.

TABLE 6-1

TYPICAL IMAGE PROCESSING SUBSYSTEM OPERATIONS		
Memory Operations	**Neighborhood Operations**	**Multi-Image Operations**
Subsample	Convolution	Add
Indirect Access	Dilation	Multiply
Move	Erosion	Subtract
Interlace	Maximum	Maximum
	Minimum	Minimum
	Logical Or	Select

the model of the Intelligent Image Memories includes the ability to fetch and/or store more than one image at a time.

The individual instructions that can be executed by a functional unit (processing element) include neighborhood and multi-image operations (Figure 6.2-3). Neighborhood operations require simultaneous access to neighborhoods of pixels in one input image. The value of each output pixel depends on several pixels from the input image. Multi-image operations require simultaneous access to corresponding pixels in more than one image. The value of each output depends on one pixel from each input image. These instructions are executed on a stream

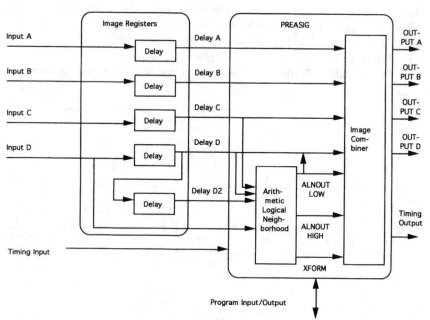

Figure 6.2-3.   Pipeline Processing Element.

of input pixels in a way that is similar to that of systolic processors. That is, each functional unit receives an instruction to execute some neighborhood and/or multi-image operation. Then an input data stream is created that, at each clock, provides all of the pixels required to produce one output pixel. These data are clocked into the functional unit simultaneously. The functional unit is pipelined to produce the required output after several clock cycles. With the internal pipelining, one set of inputs can be clocked in and one output can be clocked out of every functional unit for each clock cycle.

The IPS model includes registers associated with each functional unit for temporary storage of images before they are operated on. The registers only need to hold pixels until the functional unit is ready for them so it is not generally necessary for them to hold an entire image. Instead, each register acts as a FIFO, delaying a stream of image data until needed. The latency between when a pixel enters a register and when it leaves is determined by the operations being performed by the functional units. Also, there are two types of registers, distinguished by their number of outputs. One provides one pixel of data from an image at each clock cycle. The other is used for neighborhood operations and provides up to nine pixels of data from an image at each clock cycle.

The Image Processing Subsystem programming model also exploits the fact that the same operations are usually performed over an entire image or window within an image. The series of operations in a algorithm is broken into groups of instructions that can be executed in one pass by the Processing Elements in a pipeline. The size of the group depends on how many Processing Elements are provided. The Image Memories, Pipeline Processing Elements, and Feature Extractors are programmed to retrieve the necessary data, execute a group of instructions, and return any temporary results. Once the instructions have been loaded a trigger is generated that starts execution of that group of instructions. While the instructions are being executed the Image Processor Controller can load the next group of instructions into the Image Memories, Pipelined Processing Elements, and Feature Extractors. This process of reading data, executing one group of instructions, and returning the data to the image memories is called a *recirculation*.

The Image Processing Subsystem of ATCURE was designed to execute a wide variety of embedded real-time image interpretation programs. The architecture recognizes that there are characteristics of image manipulation functions that can be exploited to produce a high performance processor. Some of these characteristics follow.

- Images are formed from large numbers of pixels.
- Similar operations are executed on all of the pixels selected from an image.
- Image data can be stored so that they are usually accessed in some predictable order.
- Neighborhood operations are about as common as multi-image operations.

- There are often long sequences of operations that do not depend on global (image-wide) analysis of the data.

A recirculating systolic architecture has been developed that exploits these characteristics. Some of its features follow.

- A single memory access retrieves or stores an entire image, or part of an image, independent of the image size.
- The basic instructions include neighborhood operations on single images and a complete complement of operations on pairs of images.
- Each instruction performed by the functional units operates on an entire image or part of an image.
- The registers within the processor store image data.
- Some image registers provide access to up to nine pixels in a neighborhood simultaneously.
- The status register lists features of an entire image or region.

These features of the IPS model eliminate many bottlenecks found in other systems in a variety of ways. During a recirculation all operations occur synchronously. At each clock cycle one pixel from each image is fetched or stored by the Image Memories, each Processing Element receives one pixel of input from each of its operands and performs one rather complex operation on the image(s), passing one pixel of results down the line. Thus, Processing Elements do not wait to receive an entire image before they start to use it in an operation. As each pixel becomes available it is used in the next calculation.

Furthermore, during each recirculation each processor in the pipeline can complete one multi-image operation and one neighborhood operation on an entire image (on an input image/data stream). With this systolic pipeline approach, an IPS can execute the equivalent of thousands of general purpose processor instructions each clock cycle. Furthermore, because each instruction is executed over entire images or subimages, each instruction fetched is executed thousands of times.

An ATCURE system with 20 processing elements in the pipeline has been constructed. It is capable of performing 8×10^9 mathematical operations per second on images with up to 2000×2000 16-bit pixels, a capability equivalent to executing 4×10^{10} instructions per second of a typical general purpose RISC instruction set.

REFERENCES

1. Barr, Avron, and Feigenbaum, Edward A., eds., *The Handbook of Artificial Intelligence*, William Daufman, Inc., Los Altos, Vol. 1, 1981, Vols. 2 and 3, 1982.

2. Rich, Elaine, *Artificial Intelligence*, McGraw-Hill Book Co., New York, 1983.

3. Jain, Prem P. and Newton, Peter, Putting Structure into Modeling, *Unknown*, pp. 49–54, 1989.

4. Uhr, L., et al., *Evaluation of Multicomputers for Image Processing*, Academic Press, Orlando, FL, 1986.

5. Kung, H.T., Why Systolic Architectures, *Computer*, pp. 37–46, January 1982.

6. Lougheed, R.M., Advanced Image Processing Architectures for Machine Vision, *Proceedings of SPIE: Image Pattern Recognition; Algorithms, Implementations, Techniques, and Technology*, Vol. 755, pp. 35–51, January 1987.

SUGGESTED READING

Salinger, Jeremy, *Heterogeneous computer architecture for embedded real-time image interpretation*, ERIM, Ann Arbor, MI; SPIE Proceedings volume 1957, Architecture, Hardware and Forward-Looking Infrared Issues in Automatic Target Recognition, 12-13 April 1993, Orlando, FL (SPIE, Bellingham, WA).

6.3   WARP

The Warp Array systolic processor [1] is the heart of the Warp System, which provides the computational horsepower required by many vision processing, signal processing, and scientific computing applications. The Warp System is a linear array of identical powerful floating point processors called Warp Cells. Since each Warp Cell is identical and its communication with adjacent cells identical the linear array is easily configurable from 1 to N cells by simply plugging in boards. A typical system is a linear array of 10 Warp Cells each being capable of 10 MFLOPS, therefore a 10 cell system constitutes a 100 MFLOP systolic array processor. The Warp System was developed by H.T. Kung at Carnegie-Mellon University.

Each Warp Cell is fully programmable with code normally generated by an HLL compiler (W2) or, for very special applications, written by hand using W1 (Warp Cell assembly language). Instruction execution on the cell for a single W1 instruction is 200 ns. Special conditional jump control features are built into the Warp Cell to allow for data-dependent and -independent control structures to be programmed on each Warp Cell.

One of the most powerful features of the Warp Array is the high speed synchronous data routing available between adjacent Warp Cells and within a Warp Cell. The data are sent into the array from the IU (Interface Unit) to the "left-most cell" and are sent out of the array back to the IU from the last logical cell in the system. The data flow through the system need not go through the entire physical array. Certain algorithms are better mapped into a number of cells that are less than the physical number. For instance, a 128-point complex FFT requires seven passes, each pass being mapped to a cell, therefore requiring only a seven cell logical system. There is a Global Output bus shared by all cells, which is statically enabled during the execution of each algorithm determining which cell is the last cell in the logical system (enabled for Global Output). In the Warp Array there are four high speed data paths that connect adjacent cells and two Global Output buses that are used to output results. Figure 6.3-1 shows how the data can be routed between a pair of adjacent cells and how they get linked into the functional units within the cell.

There are two "forward" data paths (X-next data and Y-next data), one "reverse" data path (Y-previous), and one "forward" address path (A-next), which is a path where data memory read and write addresses get passed from cell to cell. Each of the data paths shown in Figure 6.3-1 appear to the user as 32-bit paths each capable of 20 MBytes bandwidth. There are many special features built into the hardware and software control of these data paths that help achieve maximum flexibility and bandwidth without adding programming complexities.

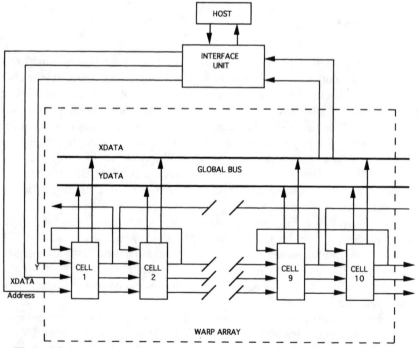

Figure 6.3-1.   Adjacent warp cell interconnection scheme.

A block diagram depicting the internal architecture of a Warp Cell is given in Figure 6.3-2.

The cell can be divided into the following four functional areas.

1. Cell data paths and features

2. Microcontrol Unit

3. Data memory organization and address generation

4. Floating point processing elements

It should be noticed that although the W2 compiler hides many of the details of the cell, more efficient W2 code can be generated if the underlying cell architecture is understood. Certain scheduling constraints and memory utilization must be adhered to when programming the Warp array.

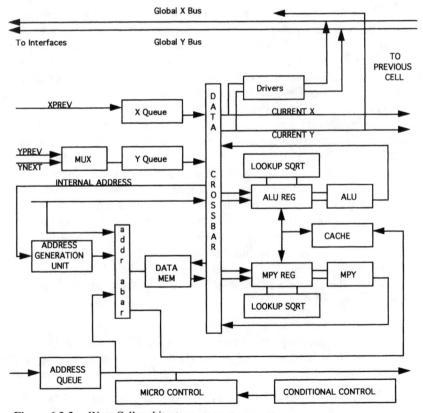

Figure 6.3-2. Warp Cell architecture.

The data flow structure of the cell is divided into two separate areas to simplify the discussions. First we discuss the cell's external I/O and flow control and then discuss the internal data routing capabilities of the cell. Data movement within the array is manipulated by "Send()" and "Receive()" statements when writing code using W2. To the programmer, the data paths (internal and external) appear as 32-bit wide paths. In actuality they are 16-bit time multiplexed data paths with the upper half of data sent in the first half of a major cycle and the lower half sent in the second half of a major cycle.

The only way data gets into a cell (excluding microcode literals) is over either the X, Y, or address paths as depicted in Figure 6.3-1. There are two different ways data leave the cell. One way is to send data to the adjacent cell via the X, Y, or address paths. The other way is to send data back to the IU over the Global X Bus. (The cell must be configured to be the last logical cell in the array.) The data leaving the cell, as mentioned earlier, can also be sent to the previous cell via the Y data path. It is important to note that the data being sent to the previous cell on the Y path is the same as the Y data being sent to the next cell. Therefore, it is possible to send the same data backward and forward simultaneously on the Y path but NOT possible to send different data to the previous cell and simultaneously to the next cell via the Y path. Normally, this does not present a problem since the user still has the forward X path available when sending Y data in the reverse direction. The same restriction exists when sending data back to the IU on the Global X bus and simultaneously sending data to the next cell on the X path. Normally this is not done since the last logical cell of the array rarely needs to send any data to its right-hand neighbor.

At the front of each input data path on the cell is a 512 word data queue. A 512 word depth was chosen to accommodate an entire row of a 512×512 standard input image. This queue acts to buffer the data streams between adjacent cells, which reduces the complexity of scheduling I/O activities by the programmer. Adjacent cells' programs need not be cycle to cycle synchronous with respect to each other when sending and receiving data. When a cell has data to send to an adjacent cell, it merely sends it without care if the receiving cell is ready for the data yet or not.

A hardware feature has been incorporated into the cell to reduce the complexity of data flow control within the Warp array. This feature involves automatic hardware intervention of program execution on the cell, which produces a data conflict. There are two different forms of data conflicts, which are described in the following.

1. Write Conflict: Condition created when a cell is sending data to a full queue (on an adjacent cell or the IU output fifo when sending on the Global X bus).

2. Read Conflict: Condition created when a cell is reading data from an empty queue (cell can only read from its own queue).

The hardware intervention is transparent to the program (and therefore the programmer) and only affects the cell causing the conflict. For a write conflict, only the cell sending to a full queue gets temporarily suspended; and remains suspended until the write conflict is resolved (adjacent cell reads a word from the full queue). Similarly, when a cell tries to read from an empty queue only that cell suspends operation until the adjacent cell sends a word to its queue, thereby resolving the read conflict. During the period the cell is suspended, the state of all the computational units within the cell are frozen until the conflict is resolved, at which point the cell continues as if no conflict had occurred.

All possible combinations of concurrent conflicts are allowed. Let's take for example a case where a cell is simultaneously sending data to the adjacent cell's X and Y queue and reading from its own X queue all in a single cycle. To further complicate things, assume the X queue on the next cell is full (causing a write conflict) the Y queue is not full (no conflict) and the cell's own X queue is empty (causing a read conflict). This case is particularly tricky because all three activities must complete as if no conflict ever happened. The microprogram will not continue to the next line of code until all the conflicts have been resolved in the current cycle. Once the last conflict has been resolved, the microsequencer quickly continues on (losing only one cycle [200 ns] for H/W overhead to get going again).

Although this hardware intervention severely reduces the complexity of scheduling the data through the Warp array, the programmer must be aware of potential S/W deadlocks.

Occasionally algorithms need to "tag" certain data elements to simplify flow control; for instance, the last word of an array, first word of an array, etc. . . . To accommodate S/W testing of tagged data an extra "TAG" bit is appended to the X and Y data streams and two extra bits appended to the address path. Therefore the X and Y data paths between cells (and back to the IU) are actually 33 bits wide and the address path is 34 bits wide. These data tags are sometimes referred to as Systolic condition code bits. Using these TAGS (which may originate from as far back as the Cluster Processors) gives each cell the ability to (under S/W control) jump on the TAGS, ignore them, pass them, alter them, or generate them. The W1 Assembler User's Manual gives a complete description of how to use and manipulate the data and address TAGS. Note that the W2 compiler utilizes the TAGS automatically and therefore does not support tag use at the W2 programming level.

The "super glue" of the internal data paths of the Warp Cell is the data crossbar. As shown previously in Figure 6.3-2, the high speed data crossbar provides a flexible "data switch" within the cell that links together all the internal elements and connects them to the cell's I/O paths. The data crossbar allows any input to go to any combination of output(s) during a major cycle (obviously no two or more inputs can be destined to the same output).

The Warp Microcontrol Unit is used to generate dynamic control to the Warp Cell. The microcode for the Warp Cell gives the user complete parallel access to the components to the cell. The microengine consists of an address sequencer, address register, microcode memory, instruction register, and conditional jump control logic. Control is generated by sequencing through locations in the microcode memory specified by the address sequencer through the address register. The data bits output from the microcode memory are registered into the instruction register that actually supply the control to all the resources within the cell. Different conditional control sequences are created by sampling a condition code (CC) bit supplied by the condition control logic.

The microcode memory is downloaded via "serial chains" prior to execution of an application session.

Although the programmer "sees" a 200-ns instruction execution time the actual microengine in the cell is operating at a 100-ns pipeline rate. The programmer's cycle is referred to as a major cycle (200 ns), which is composed of two separate back to back minor cycles (100 ns each; first the *even* minor cycle then the *odd* minor cycle). To keep the width of the microcode memory to a minimum certain fields are shared in memory and multiplexed via the instruction register every other cycle. The memory data to the instruction register is 136 bits wide and the instruction register is 208 bits wide. All of this is transparent to the user except the fact the microcode memory is 16,384 (16 K) locations deep which allows storage of 8192 (8 K) W1 level instructions (8 K even, 8 K odd).

Each Warp Cell contains a 32 K data memory (32,768 × 32 bits) which can be used for storage of intermediate results, parameter storage, input storage, and so on.

The internal structure of the data memory is 32 bits. This allows two 32-bit accesses (one read and one write) every major cycle (200 ns). The write port (WPORT) and read port (RPORT) latch the memory data as 32-bit quantities and translate them from/to the data crossbar in the normal 16-bit "upper half then lower half" fashion as previously described. Memory reads take place during the first half of a major cycle where memory writes take place in the last half of a major cycle. Therefore, if a particular data memory location is being written and read in the same W1 cycle, the old value will be read with the new value written. During the read cycle, a 15-bit address (via the address crossbar) is presented to the memory with the corresponding memory data (on the bidirectional internal bus) latched into the RPORT. This data is then available (in the next major cycle) to be sourced to the data crossbar. During a write cycle, a 15-bit address (via the address crossbar) is presented to the memory and data written from the WPORT (which is driving the bidirectional internal bus). Data sent to either the WPORT or the RPORT remain in the ports until over-written. The RPORT serves a duel function during diagnostics. It can write or read the data memory during diagnostics since it is on a serial chain.

The floating point MPY and ALU units can each perform up to 5 MFLOPS. Each unit utilizes a seven stage pipeline that can generate one result every 200 ns. The Weitek (1232/1233) chip set conforms to the requirements of the IEEE Standard 754 for single precision floating point arithmetic.

The Weitek 1066 register file is capable of storing 31×32-bits data words for the ALU and MPY floating point unit. Each floating point unit (both MPY and ALU) has its own register file unit. In every major cycle of the Warp cell operations, the register file is capable of reading two 32-bits data words from the data crossbar, sending two 32-bits data words to the floating point processors, and sending one-32-bits word to the look-up unit.

REFERENCE

1. *Warp Cell Architecture Manual in Warp User Manual*, Version 2.0, March 1988.

6.4   TERA MTA

The Tera Multithreaded Architecture (MTA) computer, the brainchild of Burton Smith and his colleagues, is the first scalable symmetric multiprocessor (SMP) system. Typical configurations will deliver three to ten times the performance of previous supercomputers in their class. Tera's performance is due to a combination of a high clock rate, multithreaded pipelines, multiple operations per instruction, and the ability to effectively use a large number of processors in a single system.

There are several distinctions between TERA's MTA and conventional vector multiprocessors and massively parallel systems. Among these are scalability and general-purpose applicability.

A system is said to be scalable when its performance can easily be increased by making the individual components larger and/or by using more of them. A basketball team cannot be scaled up very far because soon the extra players would get in each others' way, and overall performance would actually decrease. Very similar phenomena can occur in computer systems.

General-purpose applicability means that the processing speeds are available to a wide class of problems, not just ones that lend themselves to a particular architecture.

Tera's architectural approach makes both scalability and general-purpose applicability possible. Unlike the massively parallel solution, with hundreds or thousands of microprocessors hooked together in a network, the MTA system uses many processors, but with shared memory. The MTA design eliminates the time a processor has to wait for the memory, a problem called "memory latency" that limits the performance of some other systems. Thus, more processors can be added without limiting performance, and the software programming effort required is less in comparison with systems based on massively parallel architectures. Both shared memory and automatic parallelizing compilers make this ease-of-programming possible.

Tera's MTA system can use existing software without substantial modifications. UNIX or Cray Research-based programs can run on a Tera system with only normal porting activities required. There is a large investment in Cray-based software and UNIX applications, and the ability to run these applications without substantial conversion costs is an attractive feature.

The MTA is available in versions using from 16 to 256 processors. Each processor has peak performance in excess of a billion instructions per second and a billion floating point operations per second (gigaflops). Sustained, deliverable speeds of one-half-peak speeds over a wide variety of problems is being demonstrated.

	Midrange Tera
Instructions/second	64 billion
Memory (Mb)	64,000-128,000
Disk Storage (Mb)	20 million
I/O Speed (MB/second)	10,000
Concurrent Tasks	hundreds

MTA systems are shared memory systems that are architecturally scalable. The programmer is freed from data layout concerns regardless of system size. Tera's high performance multithreaded processors provide scalable latency tolerance, and an extremely high bandwidth interconnection network lets each processor access arbitrary locations in global shared memory at 2.8 gigabytes per second.

Massively parallel systems and workstation networks depend on massive locality for good performance. Applications must be partitioned to minimize communication while balancing the workload across the processors. This task is often impractical or difficult. An MTA system can accommodate intensive communication and synchronization while balancing the processor workload automatically.

Vector multiprocessors are true shared memory systems, but rely on long vectors and massive vectorization for good performance as systems increase in size. Executing scalar code, parallel or not, is seldom cost effective on these machines. MTA systems optimize both vector and scalar code, exploiting parallelism while retaining the programming ease of shared memory.

MTA systems are constructed from resource modules. Each resource module measures approximately $5 \times 7 \times 32$ inches and contains up to six resources.

- A computational processor (CP)
- An I/O processor (IOP) nominally connected to
- An I/O device via 32- or 64-bit HIPPI
- Either two or four memory units

Each resource is individually connected to a separate routing node in the system's three-dimensional toroidal interconnection network. This connection is capable of supporting data transfers to and from memory at full processor rate in both directions, as are all of the connections between the network routing nodes themselves.

The three-dimensional torus topology used in MTA systems has eight or sixteen routing nodes per resource module with the resources sparsely distributed among the nodes. In other words, there are several routing nodes per computational processor rather than the several processors per routing node that many systems employ. As a result, the bisection bandwidth of the network scales linearly with the number of processors.

Just as MTA system bandwidth scales with the number of processors, so too does its latency tolerance. The current implementation can tolerate average memory latency up to 384 cycles, representing a comfortable margin.

6.4.1   Configurations

Model	Number of Processors	Memory (GB)	Performance (Gflops)	Bisection B/W (GB/s)	I/O B/W (GB/s)
MTA 16	16 CP + 16 IOP	16–32	16	179	6
MTA 32	32 CP + 32 IOP	32–64	32	179	12
MTA 64	64 CP + 64 IOP	64–128	64	358	25
MTA 128	128 CP + 128 IOP	128–256	128	717	51
MTA 256	256 CP + 256 IOP	256–512	256	1434	102

6.4.2   Specifications

- 64-bit data, addresses, and instructions
- Up to 128 threads per processor
- Up to eight concurrent memory references per thread
- IEEE 754 floating point arithmetic
- No data caches
- 8 KB level 1 and 2 MB level 2 instruction caches
- Fortran 77, C, and C++ with customary extensions
- Fortran 90 array syntax and intrinsics
- Automatic parallelization and vectorization
- Interprocedural analysis and optimization

- Symbolic debugging of optimized parallel code
- Transparent, scalable parallel I/O
- 64-bit fast file system with variable block sizes
- 150-MB/s disk arrays of 90D360 GB capacity
- HIPPI TCP/IP connectivity to FDDI and ATM
- Multiuser support for large and small tasks
- Self-hosted system, that is, no "front end"
- Checkpoint/restart capability
- Error correction in both memory and network

The MTA is a multiuser system in which many parallel jobs of different sizes execute concurrently. Processors are multithreaded, with up to 16 jobs competing simultaneously for a processor's instruction streams. Therefore, parallel tasks can space-share even within a single processor.

The TERA MTA supports two levels of processor scheduling. The operating system scheduler manages processors at a gross level, and determines which jobs are executing at any moment. Once loaded, each job competes with other jobs for instruction streams without intervention from the operating system. The competition is guided by user-level runtime systems, one per job, that request and release streams in response to their respective jobs' parallel workloads. The runtime also works in concert with the compiler to schedule both automatically generated and user generated parallelism within a job.

The Tera MTA architecture, implements a physically shared-memory multi-processor with multistream processors, interleaved memory units, and a packet-switched interconnection network. The memory is distributed and shared, and all memory units are addressable by every processor. The network can support a request and a response from each processor to a random memory location in each clock tick.

Each processor supports 16 protection domains and 128 instruction streams. A protection domain has a memory map and a set of registers that hold stream resource limits and accounting information. Consequently, each processor can be executing 16 distinct applications in parallel. Although fewer than 16 probably suffice to attain peak processor utilization, the surplus gives the operating system flexibility in mapping applications to processors. Although two or three parallel jobs may sustain a high average parallelism executing on a processor, so may a mixture of parallel and sequential jobs.

A stream has its own register set and is hardware scheduled. On every tick of the clock, the processor logic selects a stream that is ready to execute and allows it to issue its next instruction. Since instruction interpretation is completely pipelined by the processor, and by the network and memories as well, a new instruction from a different stream may be issued in each tick without interfering with its predecessors.

Provided there are enough instruction streams in the processor so that the average instruction latency is filled with instructions from other streams, the processor is fully utilized. Analogous to the provision of protection domains, the hardware provision of 128 streams per processor is more than enough to keep the processor busy. The extra streams let running jobs fully utilize the processor while other jobs are swapping in and out, and help saturate the processor during periods of higher memory latency (perhaps because of synchronization waits or memory contention).

Several architectural features are especially useful for scheduling. A stream limit counter in each protection domain lets the operating system impose a hardware-enforced bound on the streams that can be acquired by a job. Three user level instructions dynamically acquire and release streams in the protection domain of the invoking stream: stream_reserve, stream_create, and stream_quit. The novel stream_reserve instruction facilitates automatic program parallelization of programs; it grants the right to issue some or all of the requested number of stream_create instructions, each of which in turn will activate an idle stream and pass it a program counter and a data environment. Compiler generated scheduling code can compute a division of parallel work appropriate for the number of streams actually available, and do this before those streams begin consuming processor resources. The stream_quit instruction returns a stream to the idle state.

Programs for the Tera are written in Tera C, C++, and Fortran. Parallelism can be expressed explicitly and implicitly. Explicit parallel programming is assisted through the use of *future variables* and *future statements*. A future variable describes a location that can eventually receive the result of executing a future statement, which is executed in parallel with its invoker.

The Tera MTA has full/empty bits on every word of memory that support an efficient implementation of futures. When a future statement starts, its future variable is set empty, causing subsequent reads to block. When the future statement completes, it writes its result to the future variable and sets it full, freeing those parallel activities that were waiting to read the variable. The user runtime is responsible for scheduling and executing future statements. Implicit parallelism is identified by Tera's parallelizing compiler. Most of this is loop level parallelism: parallelism obtained by concurrently executing loop iterations. Other forms exploited include block level parallelism, executing independent blocks of code concurrently, and data pipelining (achieved only when explicit synchronization is provided).

The compiler schedules resources for much of the parallelism it generates. It generates instructions to allocate and start hardware streams, to schedule the work among the streams either statically or using a self-scheduling approach, and to release the streams. When parallelism is plentiful the compiler acquires streams on multiple processors. To do this it generates calls to the user runtime (which in turn may call the operating system) to create a stream in each remote protection domain. Compiler-generated code then acquires and releases streams in each domain based on the amount of parallelism.

In summary, the user level runtime schedules resources in response to explicit parallelism; compiler generated code adaptively schedules resources wherever implicit parallelism is detected.

The software architecture of the runtime consists of a hierarchy of objects animating the processing behavior of the Tera. Many objects parallel those of the operating system.

A running user program is represented as a *task* consisting of a collection of *teams*, each of which occupies a protection domain. The protection domains are on distinct processors if this is possible. Practically speaking, it serves no purpose to put two teams of the same task on the same processor: The stream limit imposed by the operating system is large enough to enable a parallel team to more than saturate the processor.

Each team has a set of *virtual processors* (vps), each of which is a process bound to a hardware stream. Note the distinction between a stream, a hardware entity, and a vp, a software entity. Every vp has a dedicated stream, but not every stream has an associated vp. Compiler generated parallelism makes use of streams without vps. Or a stream may simply be unused. A task starts with one team and one vp by default. The runtime is responsible for acquiring more vps if there is parallelism in the program and resources are available.

The set of active tasks at any moment is under the control of the operating system. The runtimes representing each of these tasks compete with one another for resources.

Work appears in two types of queues in the runtime. New futures are spawned into the *ready pool* of the task. The ready pool is a parallel FIFO based on fetch&add. Futures that once were blocked but are now ready to run are placed in the *unblocked pool* of the team on which they blocked. Distributed unblocked pools are a convenient means to direct special work, such as a command to exit, to a particular team.

The runtime self-schedules work from the ready and unblocked pools. Vps from every team repeatedly select and run work from the pools. A vp's unblocked pool is given precedence because we would like to complete old work before starting new work. Executing work can result in allocation of additional instruction streams for compiler generated parallelism, creation of new work, and blocking of text. The vp executing work that creates new work or unblocks old work is responsible for placing that work on the ready or on the unblocked pool, respectively. If the vp blocks, it searches for other work on the unblocked and then the ready pool; that is, blocking does not lead to busy-waiting.

Some programs, especially those using divide-and-conquer algorithms, may do better with an ordering other than FIFO. A language extension touch (short for *touch-future*) assists with the effective execution of those programs. When touch is applied to a future variable, the vp stops executing its current work and executes the future statement instead, unless the future is already executing. Thus the vp executes the work on which *its* work is waiting. This policy trades a few extra instructions if the future has already started for the overhead of blocking if it has not. Using touch makes future scheduling more LIFO than FIFO in character.

In practice, the user need not explicitly touch futures in divide-and-conquer programs. The compiler is able to infer divide-and-conquest recursive structure; when it discovers the storage class of a future variable is automatic, it generates code to perform the touch. The result is breadth-first scheduling of subproblems until vp creation lags behind subproblem creation; at that point the scheduling becomes depth-first. This dynamic schedule is both efficient regarding overheads and reactive to changes in the vp workforce.

6.4.3   Scheduling Stream Resources

The runtime is responsible for dynamically increasing and decreasing the number of vps it employs to execute the user program. This is important because the average size of the ready pool may vary significantly over the lifetime of the program. Reasonably, a large ready pool merits more vps than a small ready pool, and vps acquired for periods of high parallelism should be retired during periods of low parallelism to improve efficiency.

The goals of the virtual processor strategy of the runtime mirror those of the runtime itself: An increase in the number of vps is attempted whenever it might reduce response of the program, subject to the constraint that overall throughput of the machine should not be significantly reduced. Were acquisition and release of streams instantaneous, a policy of creating a new vp for each and every future would ensure minimum response time; retirement of vps whenever pools emptied would minimize impact on throughput. But there are overheads, and the runtime cannot foresee how pool sizes will change and how much computation any piece of work will demand. Nor does it have information regarding the behavior of the other programs with which it is competing. So there is an inherent tradeoff between minimizing response time, through zealous resource acquisition with delayed release, and maximizing throughput—assuming resources lag demand—through conservative acquisition with immediate release.

To minimize the overhead involved when a vp adds work to the ready pool, the growth policy is implemented not by those vps but by a daemon mechanism that periodically assesses growth based on idleness and work available. The daemon creates twice the number of vps previously created whenever there is *any* work to be done, and no vps were idle since the last acquisition. For example, if there is initially one vp, after one period another will be created; after two periods, assuming both vps were kept busy, two more will be created; after three periods, supposing one of the four vps found nothing to do, even for a moment, no vps will be created; and after four periods, assuming all four vps were kept busy this time, one more vp will be created. If an acquisition request fails, in whole or in part, it is repeated after the next period has elapsed.

The total overhead is at most a small constant fraction of the work intrinsic to the program, and the response time is similarly bounded from above by at most a small constant factor multiplied by the minimal response time, were all streams requested actually available. (For response time, an additive lower order term proportional to the logarithm of the maximum parallelism persists in our scheme, because of the delay of exponential growth versus instantaneous growth.) Such worst-case bounds should be achieved only when the parallelism is highly erratic in just the wrong way.

Just as important as the acquisition policy is the retirement policy: Idle vps are released according to a schedule that inverts the acquisition schedule. That is, were a large number of vps to become idle at once, one would disappear after one period, two more after a second period, and so on unless work appeared to interrupt the process. This is necessary to avoid the anomalous behavior that would result were there to be brief sequential periods separating slightly longer but highly parallel periods at just the right frequency to cause repeated rapid growth and collapse. Were idle streams released after a constant period of time independent of their number, the constant factor in the response time bound becomes logarithmic in the maximum parallelism.

Because idle vps on the Tera consume as much processing power as vps performing memory-bound computations, the TERA implementation never really allows more than one vp per team to be idle concurrently: Within a period, additional idle vps are killed, but remembered, and replaced the instant work appears. Creation and retirement within a team is through user level instructions, and so is fast enough to justify such a reactionary policy.

6.5   IMAGE UNDERSTANDING ARCHITECTURE

Parallel processing is now generally accepted as necessary to support real-time image understanding applications. Much debate remains, however, about what form of parallelism to employ. Part of this debate stems from the tremendous amount and variety of potential parallelism in machine vision.

The sensory data alone is a good example; a medium-resolution image (512 $\times$ 512 pixels) consists of roughly a quarter of a million data values. In many cases, each of these values might be processed in parallel. Further, if images are obtained from a video camera, the steady stream of data lends itself to pipelined parallelism. Some data involves multiple sensors (e.g., stereo or nonvisual spectral bands), thus providing yet another potential source of parallelism. Nor is it unusual to extract many different features from a given image or set of images (e.g., lines, regions, texture patches, depth maps, and motion parameters), and these processes may also be carried out in parallel.

Beyond the sensory data, image understanding involves knowledge-based processing; and between these two levels of abstraction, symbolic processing has proved useful. Thus, vision researchers tend to classify algorithms and representations into three levels: low (sensory), intermediate (symbolic), and high (knowledge-based).

Of course, the existence of multiple levels of abstraction is yet another source of potential parallelism. Moreover, processing within each level presents many possibilities for exploiting parallelism. Part of the allure of developing a vision machine, from a computer architect's perspective, is this tremendous quantity, diversity, and complexity of latent parallelism. By comparison, most scientific and engineering applications have simple organizations with straightforward requirements for parallelism. (For more detailed analysis of the potential for parallelism in image understanding, see Weems [1].)

Over the past few years, the University of Massachusetts and Hughes Research Laboratories have worked together to develop a hardware architecture that addresses at least part of the potential parallelism in each of the three levels of vision abstraction. A 1/64th-scale proof-of-concept prototype of this machine has been built. The machine, called the Image Understanding Architecture (IUA), consists of three different, tightly coupled parallel processors; the content addressable array parallel processor (CAAPP) at the low level, the intermediate communication associative processor (ICAP) at the intermediate level, and the symbolic processing array (SPA) at the high level. Figure 6.5-1 shows an overview of the architecture. The CAAPP and ICAP levels are controlled by an array control unit (ACU) that takes its directions from the SPA level.

The SPA is a multiple-instruction multiple-data (MIMD) parallel processor, whereas the intermediate and low levels operate in multiple modes. The CAAPP operates in single-instruction multiple-data (SIMD) associative or multiassociative mode, and the ICAP operates in single-program multiple-data (SPMD) or MIMD mode. In multiassociative mode, CAAPP cells execute the same instruction stream but in disjoint groups, with each group capable of operating on locally broadcast values and locally computing its own summary values in parallel with other groups. In SPMD mode, the ICAP processors execute the same program but have their own instruction pointers so that they can branch independently.

The IUA addresses the various forms of potential parallelism described in the introduction by using the capabilities of each level in Figure 6.5-1. The I/O staging memory permits one or more sensors to input images into a buffer that can hold up to 15 seconds of imagery at 30 frames per second and a resolution equal to the size of the low-level processor array. The resolution of the images can differ from the array size, with a resulting increase or decrease in the number of frames that can be buffered.

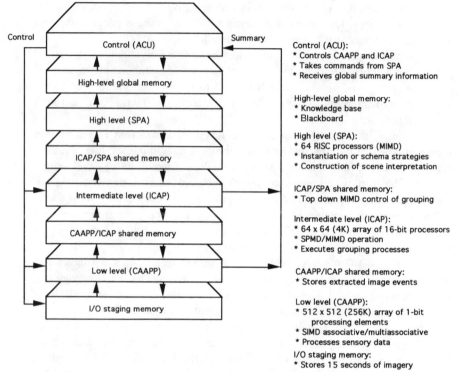

Figure 6.5-1.   Overview of first-generation Image Understanding Architecture.

The CAAPP consists of bit-serial processors, each with an arithmetic logic unit, registers, 320 bits of explicitly managed on-chip cache memory, and 32 Kbits of backing store (main) memory. Because it is a SIMD processor, its instructions are broadcast from the ACU. However, each processor also contains a one-bit register that controls whether it will respond to a particular instruction.

The processors are connected via a reconfigurable mesh, called the *coterie network*. Each processor controls four switches that configure the mesh connections to its four nearest neighbors (north, south, east, west) and four switches that permit signals to bypass the processor (northeast, northwest, horizontal, and vertical). When the switches are set, connected processors form a coterie. The mesh may simultaneously contain many nonoverlapping coteries. The coterie network connects the coteries.

Within a coterie, one processor may be selected to broadcast a value to the members of the coterie, or any subset of the processors may send a value bit-serially over the network. In the latter case, the processors receive the logical OR of the bits that were transmitted—that is, if some of the processors transmit a 1, then all processors receive a 1; however, if none of the processors transmits a 1, then all processors receive 0. This some/none test is a valuable summary mechanism that can be used in many ways. For example, it can be used to determine the maximum of a set of values contained in a coterie.

If the array has been split into coteries corresponding to regions in an image, then we can use the maximum-value operation to label connected components. Each processor is merely given a unique value (its address) and then the maximum-value operation determines the maximum address within each coterie. The value is then broadcast to the members of the coterie as their component label. Note that all of this takes place in every coterie simultaneously, even though there is only a single instruction stream. In the CAAPP, connected-components labeling thus takes only about 50 microseconds. Many other operations on image regions and edges can be performed quickly when the network is arranged to match their shape. The ability to simultaneously perform queries and summarize results in independent groups of processors under a single instruction stream resulted in the term multiassociative for this mode of parallelism.

The main memory for the CAAPP is also directly accessible to the ICAP through a second port. Each ICAP processor has access to the 8×8 tile of CAAPP processors below it, providing a highly parallel data path between the two levels. Each ICAP processor is a 16-bit digital signal processor (DSP) with 128 Kbytes of program memory and 128 Kbytes of data memory. A DSP was selected because it provides a set of operations (such as single-cycle square and add) that are well suited to computations in spatial geometry. The DSP is also designed for use with a minimum amount of external logic, and it provides a set of communication channels that are used for interprocessor communication. As an example of its capabilities, the intermediate level can simultaneously match several thousand models against symbolic descriptions of events (tokens) extracted from an image by the CAAPP.

The ICAP connects to another dual-ported memory, which it shares with the SPA. Each SPA processor can access data stored in this memory by any ICAP processor. A commercially available multiprocessor is used at this level to provide general-purpose computational capabilities for high-level processing. The SPA also has its own shared memory. The ACU, which manages the CAAPP and ICAP, is connected to that memory and communicates with the SPA processors as if it were just another processor of the same type. The full-scale IUA can thus process in parallel all pixels of a single 512×512 image, several thousand tokens, and up to 64 high-level processes. Simulations of the full-scale IUA have shown that it can support model-based recognition tasks at or near frame rate, which is considerably closer to real-time image understanding than previous systems. Nonetheless, even greater parallelism will be required to achieve true machine perception. (For more information on the IUA, see Weems et al. [2].)

Knowledge-based machine vision is both complex and computationally intense. It also presents a unique set of opportunities for exploiting parallelism. The

Image Understanding Architecture has been built to capitalize on several of those sources of potential parallelism. Because the capacity for complex parallelism in vision is far beyond the capabilities of current technology, parallel architectures for vision will continue to evolve at the forefront in architectural research [3].

REFERENCES

1. Weems, C.C., The Architectural Requirements of Image Understanding with Respect to Parallel Processing, *Proc. IEEE*, Vol. 79, No. 4, Apr. 1991, pp. 537–547.

2. Weems, C.C., et al., The Image Understanding Architecture, *Int'l. J. Computer Vision*, Vol. 2, No. 1, Jan. 1989, pp. 251–282.

3. Weems, Charles C., Riseman, Edward M., and Hanson, Allen, Image Understanding Architecture: Exploiting Potential Parallelism in Machine Vision, University of Massachusetts, Amhurst, MA; *IEEE Computer,* February 1992, p. 65.

Chapter 7

Distributed Heterogeneous Supercomputing

Some applications admit the use of several supercomputers at once, particularly when various parts of the program match best with different machine architectures. For example, a given application may involve a section of code for image processing that matches well with an SIMD machine, whereas the other parts of the code are poorly matched to SIMD but well matched to an MIMD machine or a neural net machine, or even a vector processor. In these cases a distributed heterogeneous software environment fosters efficient use of the available resources.

Four of these distributed heterogeneous software environments are described in this chapter: Parallel Virtual Machine (PVM), available free from the Oak Ridge National Lab; the GE Distributed Application Environment, developed by the GE Advanced Technology Laboratories, now a part of the Lockheed Martin Corporation; Linda, commercially available from Scientific Computing Associates; and ISIS, available from Cornell University.

7.1   PARALLEL VIRTUAL MACHINE (PVM)

Wide-area computer networks have become a basic part of the computing infrastructure of today. These networks connect a variety of machines, representing an enormous computational resource. This section describes two software packages to facilitate the use of a network of heterogeneous computers. The first, Parallel Virtual Machine (PVM), allows utilization of the network of machines as a single computational resource. The second, Heterogeneous Network Computing Environment (HeNCE), is a graphic tool to assist the user in writing and analyzing parallel programs. This section describes the parallelization of a large materials science application code, including modifications required to run on heterogeneous supercomputers.

PVM is a software package that enables concurrent computing on loosely coupled networks of processing elements. PVM may be implemented on a hardware base consisting of different machine architectures, including single CPU systems, vector machines, and multiprocessors. These computing elements may be interconnected by one or more networks, which may themselves be different (e.g., Ethernet, the Internet, and fiber optic networks). These computing elements are accessed by applications via a library of standard interface routines. These routines allow the initiation and termination of processes across the network as well as communication and synchronization between processes.

Application programs are composed of *components* that are subtasks at a moderately large level of granularity. During execution, multiple *instances* of each component may be initiated. Figure 7.1-1 depicts a simplified architectural overview of the PVM system.

Application programs view PVM as a general and flexible parallel computing resource that supports an MP model of computation. This resource may be accessed at three different levels: (1) the *transparent* mode, in which component instances are automatically located at the most appropriate sites; (2) the *architecture-dependent* mode, in which the user may indicate specific architectures on which particular components are to execute; and (3) the *low-level* mode, in which a particular machine may be specified. Such layering permits flexibility while retaining the ability to exploit particular strengths of individual machines on the network. The PVM user interface is strongly typed; support for operating in a heterogeneous environment is provided in the form of special constructs that selectively perform machine-dependent data conversions. All communications done between PVM processes use the external data representation standard (XDR) [1]. Interinstance communication constructs include those for the ex-

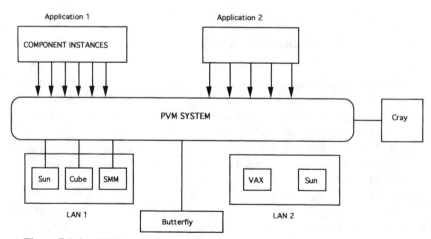

Figure 7.1-1. PVM architecture model.

change of data structures as well as high-level primitives such as broadcast, barrier synchronization, and rendezvous.

Application programs under PVM may possess arbitrary control and dependency structures. In other words, at any point in the execution of a concurrent application, the processes in existence may have arbitrary relationships between each other, and, further, any process may communicate and/or synchronize with any other. This is the most unstructured form of crowd computation, but in practice, a significant number of concurrent applications are more structured. Two typical structures are the trees and the "regular" crowd structure. We use the latter term to denote crowd computations in which each process is identical; frequently such applications also exhibit regular communication and synchronization patterns. Any specific control and dependency structure may be implemented under the PVM system by appropriate use of PVM constructs and host language control flow statements.

PVM is available through *netlib*. To obtain a copy, send mail to *netlib@ornl.gov* with the message: *send index from pvm*.

Although PVM provides low-level tools for implementing parallel programs, HeNCE provides the programmer with a higher level abstraction for specifying parallelism. The HeNCE philosophy of parallel programming is to have the programmer explicitly specify the parallelism of a computation and to aid the programmer as much as possible in the tasks of designing, compiling, executing, debugging, and analyzing the parallel computation. Central to HeNCE is a graphical interface that the programmer uses to perform these tasks. This tool supports visual representations of many of its functions.

In HeNCE, the programmer is responsible for explicitly specifying parallelism by drawing graphs that express the parallelism. The user directly inputs the graph using tools that are part of the HeNCE environment. Each node in a HeNCE graph represents a procedure written in either Fortran or C. There are three types of arcs in the HeNCE graph. One represents dependency; a second represents looping; and the third represents pipelined sections. An arc from one node to another represents the fact that the tail node of the arc must run before the node at the head of the arc.

Once the graph is complete, HeNCE will automatically write the parallel program, including all the communication and synchronization routines using PVM calls. HeNCE tools assist the user in compiling this program for a heterogeneous environment.

Graphical tools in HeNCE allow the user to dynamically configure a parallel collection of machines into a parallel virtual computer. During execution, HeNCE dynamically maps procedures to the machines in the heterogeneous network based on a user-defined cost matrix. HeNCE also collects trace and scheduling information, which can be displayed in real time or saved to be replayed later.

During the last few years, ORNL material scientists and their collaborators at the University of Cincinnati, SERC at Daresbury, and the University of

Bristol have been developing an algorithm for studying the physical properties of complex substitutionally disordered materials. A few important examples of physical systems and situations in which substitutional disorder plays a critical role in determining material properties include metallic alloys, high-temperature superconductors, magnetic phase transitions, and metal/insulator transitions. The algorithm being developed is an implementation of the Korringa, Kohn, and Rostoker coherent potential approximation (KRR-CPA) method for calculating the electronic properties, energetics, and other ground-state properties of substitutionally disordered alloys [2]. The implementation allows the treatment of materials having complex atomic lattices and having any number of disordered sublattices. The algorithm also solves the underlying quantum mechanical equations within a fully relativistic framework [3]. This is necessary for materials containing heavy elements such as barium in which even the outer, bonding electrons are traveling at a significant fraction of the velocity of light.

The KRR-CPA method extends the usual implementation of local density approximation to density functional theory (LDA-DFT) [4] to substitutionally disordered materials [5]. In this sense it is a completely first-principles theory of the properties of substitutionally disordered materials, requiring as input only the atomic numbers of the species making up the solid.

The KKR-CPA algorithm contains several locations where parallelism can be exploited. These locations correspond to integrations in the KKR-CPA algorithm. Evaluating integrals typically involves the independent evaluation of a function at different locations and the merging of these data into a final value. The two most obvious locations for parallelization in the KKR-CPA algorithm are in the integration over the Brillouin zone and the integration over energy. Each location was evaluated in terms of the available parallelism and the required communication overhead incurred by splitting the algorithm at that point. The Brillouin zone integration is the main step in each CPA iteration. The disadvantage of parallelizing the Brillouin zone integration is the large amount of communication volume that would be required. For this reason the integration over energy was parallelized. Typically, this integration involves the evaluation of the single-site Green's function for between 200 and 1000 energies in order to determine the charge density for the next self-consistency iteration. Each of these tasks involves the iterative solution of the CPA equations for the given energy and requires significantly more computation than communication.

The parallel implementation is based on a master/slave paradigm to reduce memory requirements and synchronization overhead. In this implementation, one processor is responsible for reading the main input file, which contains the number of processors to be used, the problem description, and the location of relevant data files. This master processor also manages the LDA-DFT charge self-consistency iteration. The slave processors require only enough memory to solve the CPA equations for a single energy, which presently requires 7 Mbytes. Memory reduction was a crucial consideration for porting the KKR-CPA code to the Intel iPSC/860, because the individual nodes only contain 8 Mbytes of RAM. If only one processor is requested, then a subroutine is called that calculates the tasks serially one after another. When more than one processor is requested, a pool-of-tasks scheme is used to accomplish dynamic load balancing. In this scheme, the tasks are arranged in a queue in approximate order of decreasing difficulty and assigned to idle slave processors as they become available. Thus, all processors are busy as long as there are tasks in the queue.

Additional modifications were required to allow this KKR-CPA code to execute on multiple heterogeneous supercomputers. On a Cray YMP, two options are now available in the code. The user can specify a number of tasks to initiate simultaneously on the Cray multiprocessor (ideally one per processor). A second option is to set a Cray multitasking switch in the input file and then initiate just one multitasking task. To use an iPSC/860 an intermediate routine was written that runs on the Intel host and performs three functions. This routine allocates the number of node processors specified by the master process. It receives messages from other PVM processes, converts these messages to Intel node messages, then passes the messages to the appropriate nodes (and vice versa). Finally, it releases the nodes when the problem is finished. Because of the size of the KKR-CPA code (16,000 lines of Fortran 66), we have not ported the code to run on Thinking Machines CM2, which requires applications to be written in Fortran 90.

So far, the results on a network of heterogeneous supercomputers show the viability of using such a system. Test programs of simple tasks like solving the Mandelbrot problem have been run on a DECStation, Cray XMP, iPSC/860, and CM2 configuration.

The KKR-CPA code was initially run on a network of IBM RS/6000s. Using six model 530s and four model 320s, an execution rate of 207 MFLOPS was measured. This performance is comparable to running it on a single processor of a Cray YMP.

The KKR-CPA code was also run on a virtual machine consisting of an RS/6000, a Cray XMP, and 16 nodes of the iPSC/860. Performance was not measured during these initial runs, which were designed to test the viability of the system. Performance tests will be done when a dedicated Cray YMP/8 and 128-node iPSC/860 are simultaneously available. Experiments are now underway that involve running the KKR-CPA code on a virtual machine composed of three dedicated Cray YMPs.

These experiments are the initial steps in showing how several Cray and Intel supercomputers can be configured into a virtual super-parallel computer in order to achieve the computational power necessary to solve Computational Grand Challenge Problems.

REFERENCES

1. SUN Network Programming Manual, Part Two: Protocol Specification, 1988. Beguelin, Adam and Dongarra, Jack, Solving Computational Grand Challenges Using a Network of Heterogeneous Supercomputers, Mathematic Sciences Section, Oak Ridge National Laboratory, Oak Ridge, TN and University of Tennessee; Geist, Al, Oak Ridge National Laboratory; Manchek, Robert, University of Tennessee, Computer Science Dept., Knoxville, TN; Sunderam, Vaidy, Math and Computer Science Dept., Emory University, Atlanta, GA.

2. Stocks, G.M., Temmerman, W.M., and Györffy, B.L., Complete solution of the Korringa-Kohn-Rostoker coherent potential approximation: Cu-Ni alloys. *Phys. Rev. Lett.,* Vol. 41, p. 339, 1978.

3. Ginatempo, B. and Staunton, J.B., The electronic structure of disordered alloys containing heavy elements—an improved calculational method illustrated by a study of CuAg alloy. *J. Phys. F,* Vol. 18, p. 1827, 1988.

4. von Barth, U., Density functional theory for solids, in *Electronic Structure of Complex Systems,* Phariseau and Temmerman, eds. NATO ASI Series, Plenum Press, New York, 1984.

5. Johnson, D.D., Nicholson, D.M., Pincki, F.J., Györffy, B.L., and Stocks, G.M., Total energy and pressure calculations for random substitutional alloys. *Phys. Rev. B,* Vol. 41, p. 9701, 1990.

7.2   GE DISTRIBUTED APPLICATION ENVIRONMENT (GEDAE)

7.2.1   Introduction

The General Electric Distributed Application Environment (GEDAE) is a software system developed at the Advanced Technology Laboratories (Moorestown, NJ). It provides to software developers a tool that facilitates object oriented application code generation and execution on systems ranging from single workstations to distributed heterogeneous networks.

As systems become more complex, it will become necessary to investigate solutions that incorporate a heterogeneous mix of processors interconnected on a network. The processing required by different portions of the overall system may be best suited to processors that have widely different architectures optimized for different forms of computing. This type of hardware solution is commonly referred to as an open architecture. An example of such a network is shown in Figure 7.2-1.

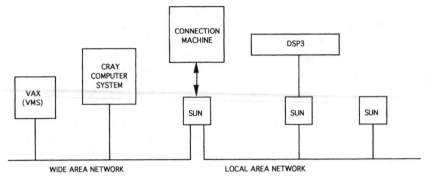

Figure 7.2-1.   Open architecture network.

This evolution has made the job of the applications software developer more complicated and hence more difficult. Each time an application is developed the programmer must deal with the differences among the machines in the distributed system, differences such as word size, protocols, operating systems, data structures, languages, buffer sizes, and so on.

Some of this complexity is being alleviated by the emerging distributed operating systems such as Mach, Alpha, Cronus, as well as Posix and OSF/DCE. However, a need persists for a programming environment that avoids the necessity to deal with interface details again and again, particularly when the processors attached to the LAN include parallel processors as well as vector processors such as Cray and Convex. The environment should present the application programmer with a common interface to all machines on the network and should hide from the developer all of the complex infrastructure of both the network and the software development environment; GEDAE provides such a development environment.

GEDAE was designed: (1) to be compliant with the Motif styleguide; (2) to support function migration between processors; (3) to operate under Vax/VMS as well as Ultrix and Unix; (4) to improve the efficiency in anticipation of wide area network operation; and (5) to allow both event driven and data driven processing.

The central concept around which GEDAE is designed is the object oriented paradigm for programming. Derived from the artificial intelligence community, object oriented programming is today finding widespread use in all areas of software engineering including database management and simulation. The basic idea is rather straightforward; software objects are self contained packages of private data, private procedures, perhaps other objects, and an interface that allows interaction with other objects. Programming in the object oriented context is directing the interactions of these objects. Object oriented programming has become popular because objects are often unaffected by software changes elsewhere in a program and are sometimes reusable in other programs, so life cycle costs tend to be reduced. GEDAE extends the object oriented paradigm to a distributed system in which objects can exist on any processor on the network which supports GEDAE.

GEDAE is comprised of four layers, namely, a Flow Graph Editor (FGE), the Distributed Application Shell (DAS), a Distributed Object System (GEDOS), and a Communications layer, as shown in Figure 7.2-2. The Application Software Libraries are application modules associated with each of the processors on the network. These modules "plug in" to the DAS, which acts as a software backplane for interconnecting the modules regardless of where they reside in the system. It is important to note that only the Application Libraries need to be developed for a given application unless processors are added to the network for which the GEDAE kernel has not been ported.

The communication layer encapsulates the operating-system-specific communication protocols, enabling GEDAE to be easily ported to various computers, including high performance computers, and to grow to take advantage of advances in distributed operating systems, such as Mach and POSIX, as they become available.

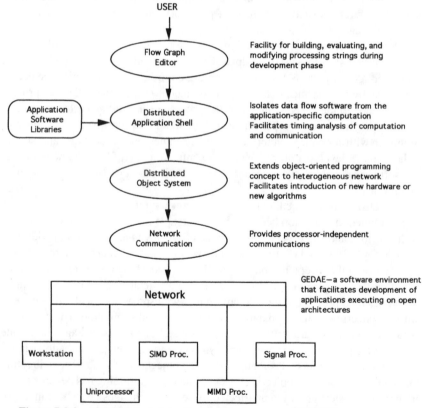

Figure 7.2-2.   GE Distributed Application Environment (GEDAE).

The communication layer gives higher layers the ability to start and connect processes in different processors on the network, to transfer data of a variety of types between processors having different internal data formats, and to send asynchronous, interrupting commands between processors.

The communication layer is small and easy to port. Currently, porting only requires the ability to create TCP/IP domain sockets. The original communication layer was written for Sun Unix and has since been ported to Vax VMS and Vax Ultrix operating systems. This experience, coupled with the small size of the communication layer (less than 2000 lines of C code), supports the contention that it can be readily ported to different operating systems.

The GE Distributed Object System (GEDOS) provides an object-oriented development environment in which a common set of functions operate on objects transparent to their location in a network of processors. GEDOS is constructed as a functional extension of the C language. It provides such functions as creating new objects, adding methods to an object, and sending messages to an object. GEDOS uses standard single-class inheritance, which provides a class hierarchy. GEDOS also allows objects to maintain a list of child objects; thus, it provides an aggregation capability.

The GEDOS layer provides all of the normal benefits of object-oriented programming to the layers above. Many of these benefits help the application developer directly. In addition to the normal benefits of object-oriented development, applications developed in GEDOS can be written for a single processor but will work just as well on a distributed network of processors. Thus, GEDOS hides the network from the layers above. Another benefit is that GEDOS objects created on one processor can be moved easily to another processor. This capability is used to map different functions (encapsulated as objects) to different processors, which can be done before or during the execution of the application.

The Distributed Application Shell (DAS) provides the data-flow control software that is used by each module in the system. It provides the interfaces to the Flow Graph Editor through which application modules can be added and connected together in a data flow graph. It also provides a simple interface for application programmers to incorporate new function modules. Any module can be divided into two parts, namely, the algorithm and the data flow specification. The Algorithm is a function written independent of GEDAE that performs the data transformation processing required by the module. The GEDAE Data Flow Specification is the data flow requirements of a module that are specified by providing simple wrapping functions that call the algorithm. These wrapping functions interface to GEDAE through standard data-flow-requirement specification functions provided by the DAS.

The Flow Graph Editor provides the user with the ability to create, display, inspect, and control the entire application through a menu-driven (Motif Standard) interface. The FGE allows a user to program a process by manipulating on a display screen the boxes of a flow diagram of that process.

Display functions can be inserted at any point in the flow diagram and dynamically change during the execution of the process. Individual function boxes in a flowgraph can be moved from one processor to another and the timing capabilities used to aid in the load balancing. This does not suggest that GEDAE provides automatic remapping of function boxes from one machine architecture to another but rather suggests that if equivalent function boxes exist for multiple machines, the specific machine used can be simply changed via flowgraph interaction.

The GEDAE software structure, as discussed in the preceding text, allows an application programmer to develop modules independently of how they are connected; and it allows a user to connect modules independently of how they are implemented and where they are executed. Data transfers across the network, data casting (translation) between different data types, and data transfers between host and remote processors are all transparent to the user. The function box application programmer can develop a module without knowing how it will be used in the system because he does not need to specify how data is transferred. This knowledge is embedded in the input and output data objects themselves. Many standard data types are provided by GEDAE, and tools for creating new data types are available.

A variety of applications have been implemented in the GEDAE environment. The first, and perhaps most comprehensive is a communication link simulation that includes transmitter, channel model, and receiver. The purpose of the simulation was to construct and optimize a modem design. This application can be demonstrated both on a single workstation and on a network of workstations. This application was demonstrated on a network in which some of the functions are executing on Sun workstations under Unix, others on a Vax operating under Ultrix and still others operating on a Vax operating under VMS.

A complete video tracker has been simulated in GEDAE. This is an image processing application in which video frames are passed through a tracking algorithm for the purpose of tracking a target within the field of view. The tracker contains a normal track mode, coast mode, and a search and reacquisition mode.

Earth science applications have also been implemented in GEDAE. These have ranged from image file browsing (simultaneously at low and high resolution) to the decoding and processing of Advanced Very High Resolution Radiometer (AVHRR) and Coastal Zone Color Scanner (CZCS) data collected from orbiting satellites.

7.3   LINDA

Multiprocessor systems can now house more power than traditional supercomputers at a fraction of the cost. Yet, without the right software tools, programs cannot easily take advantage of the parallel hardware. If there was a simple way to translate a sequential program into a parallel program, the parallel hardware use could be maximized. Linda, a parallel-programming methodology, is designed to do this.

Even with the latest parallel hardware, parallel processing is left unexploited in areas where it could be helpful. One barrier to parallel applications is the hardware-specific nature of most parallel programming languages.

Parallel computers can usually be broken into two primary architectural groups: shared memory and message passing (MP). With traditional parallel-programming methods, a program written for a shared-memory system would differ considerably from the same program written for an MP system. Because it has been implemented on both MP and shared-memory architectures, Linda has the potential to become a truly portable standard for parallel programming.

With Linda [1], parallel programs can be written in conventional programming languages, such as C, Fortran, and Lisp by supplying—in a language-independent manner—primitive operations by interprocess communication. The advantage of adding these operations to an existing language is that it creates a parallel dialect of the language.

At the heart of this mechanism is a tuple space—an abstract object through which programs place and remove objects called tuples. A tuple is a collection of related data, containing a key that retrieves the tuple.

There are four fundamental Linda operations on a tuple space: "out", "in", "eval", and "rd". For example, the tuple

```
("hello", 5, true);
```

contains three items; a string, an integer, and a Boolean value. The operation

```
out("hello", 5, true);
```

places the tuple into the tuple space. The "out" operation never blocks, nor does it return from the "in" operation, and there can be any number of copies of the same tuple in the tuple space.

Tuples are removed from the tuple space by matching them against a template. For example, the operation

```
int i; boolean b;
in("hello",?i, ?b);
```

will match any tuple whose first element is the string "hello", second element is an integer, and third element is a Boolean value. If a matching tuple is found, the variables "i" and "b" are assigned the corresponding values from the matching tuple and the tuple is removed from the tuple space. If no match is found, the process that makes the "in" call blocks until a matching tuple becomes available.

If the template matches more than one tuple, an arbitrary one is picked. The "rd" operation is similar to the "in" operation except that the matching tuple is not removed from the tuple space. "Inp" and "rdp" are variations of "in" and "rd" that do not block, even if a matching tuple is not available.

The "eval" operation is similar to "out" except that it creates an active tuple. For example, if foo is a function that returns an integer, then

```
eval("hello", foo(z), true);
```

creates a new process to evaluate foo. It then proceeds concurrently with the process that made the "eval" call. When foo returns, it leaves a passive data tuple in the tuple space, identical to an "out". These Linda operations will be used in the following.

Most sequential computer programs contain some inherent parallelism that can be exploited for faster execution on parallel computers. When analyzing a sequential program, one must first look at the structure of the results and ask if the result can be broken into smaller pieces that can be computed separately, such as the elements of a matrix or array. Second, examine the dependencies between the elements of the result. In the best case, the elements can be computed independently, yielding a high degree of parallelism. Even if dependencies exist, computation may still take on a parallel form, relying on synchronization code to coordinate the dependencies. Finally, one must examine the sequential code for nested loops, which often indicate computational chunks that could be processed in parallel rather than sequentially.

Much research has gone into parallelizing compilers, which are compilers that automatically transform sequential code into parallel code. However, a growing consensus agrees that to achieve the highest performance level, it is necessary to recode programs to use explicitly parallel algorithms, rather than extract the partial parallelisms of sequential programs.

Parallel solutions always involve a trade-off between communication and computation, which are characterized by their granularity. A fine-grain solution creates many small tasks, whereas a coarse-grain solution creates a few large ones. If the task is too small, the processes spend too much time communicating results and getting new tasks. If it is too big, then some processes might sit idle waiting for others to finish. Therefore, the programs must be carefully balanced. Linda makes it easy to adjust these factors by allowing users to alter the size of the task.

Parallel programs must be designed to minimize the synchronization delays created when one process is required to wait for the result of another before proceeding. Most problems in the real world involve data dependencies requiring synchronization processes.

Creating a parallel program, either from scratch or from an existing sequential program, includes decomposing the tasks into pieces that can be computed concurrently. There are two main types of parallel problem decomposition: data and functional. Both can easily be implemented in Linda. The type chosen depends on the nature of the computational problem. The two categories actually overlap somewhat; a parallel solution may incorporate aspects of both methods.

In data decomposition, data are divided into parts and multiple copies of an operation are applied to different parts of the data concurrently. This model is appropriate when the solution to a problem is seen as applying a process repeatedly to a collection of data objects. For example, when performing a smoothing algorithm on an image, it would be possible to divide the image into rectangular regions and have multiple smoothing processes computing in parallel.

The parts could then be recombined into the resulting image. Data decomposition is typically implemented as one master process that divides the data and multiple worker processes, which perform the computation Figure 7.3-1. The master-worker model is also referred to as a "processor farm".

In Linda, all communication between the master and workers is done through the tuple space. A tuple space is an abstract object through which programs place and remove objects called tuples. As indicated earlier, a tuple is a collection of related data containing a key that retrieves the tuple. The master divides the data and packages them as tuples that are "outed" (placed in a tuple in the tuple space) as tasks.

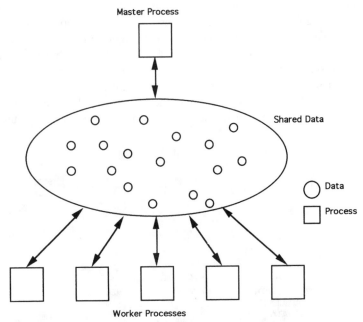

Figure 7.3-1.   One large process can be broken up into many smaller processes to help divide the workload to perform a computation.

The workers retrieve the tasks from the tuple space with "in" or "inp" operations (operations that remove a tuple from the tuple space), compute the results, and then "out" the results as tuples. Meanwhile, the master is "ining" the results as they become available from the workers. Linda also synchronizes because a process will block if it tries to "in" a tuple that is not yet in the tuple space.

A collection of similar tasks is often called a bag, because workers withdraw tasks relatively indiscriminately. This model is quite flexible because each worker continues to retrieve tasks until all are consumed—it works with any number of worker processes. The program could be written and tested on a system with one processor, using one or two worker processes.

Once debugged, it could be moved to a multiple-processor system, making it possible for more workers to take advantage of the extra processors. This model will also automatically balance the processing load among the processes. For example, one process could perform many smaller tasks while another is executing one larger task.

Functional decomposition applies different processes to a block of data. It is often the best method when a solution can be modeled as a network of nodes, such as a system simulation. Each node can be represented as a routine that processes some data and sends the results to another node for further computation. Functional decomposition usually involves some sort of explicit message passing between the processes for synchronization purposes Figure 7.3-2.

Functionally decomposed solutions often take the form of a sequential pipeline. In the case of a pipeline, a series of processes is applied to incoming data, with the results of each process passed as input to the next process in the pipe. To achieve more concurrency, small chunks of data are continually fed into the pipe so that as each process finishes its computation, more data is waiting. The danger in this structure is that the process with the heaviest computational load becomes a bottleneck, slowing all of the processes in the pipe. To avoid this, more processes must be allocated to the heaviest computation task Figure 7.3-3.

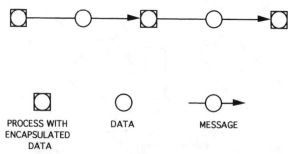

PROCESS WITH   DATA   MESSAGE
ENCAPSULATED
DATA

Figure 7.3-2.   A block of data is sent along a string of nodes, whereby each node
                does a portion of the processing.

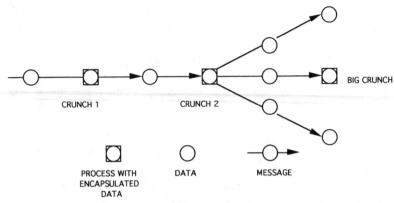

Figure 7.3-3.   If one computational load requires an abundance of processing, that load is allocated to more processes.

Linda supports functional decomposition and message passing by allowing messages to form as tuples with the destination node name as one of the keys. A node then waits for messages by "ining" on its node name. In an example with individual processes named crunch1 and crunch2, and three processes named bigCrunch, crunch2 finishes processing a block of data and "outs" the result with the Linda operation.

```
out ("bigCrunch", theResults)
```

Meanwhile, all three bigCrunch processes would be waiting for messages from crunch2 by "ining":

```
in ("bigCrunch",?theData)
```

Each bigCrunch process withdraws a message tuple as it becomes available, then computes the result, and "outs" it to the next stage in the pipe. Because of the time-intensive nature of bigCrunch, one crunch2 process could keep three bigCrunches supplied with data.

The MP model becomes inefficient when an abundance of traffic is needed to support the computation. Supplementing the MP model with some shared-data access can ease the communication load and increase the efficiency of the program.

Consider the problem of mapping a function onto a two-dimensional matrix of values. The results of the function at each location in the matrix produce a topology of the function. The third dimension of the topology is derived from the resulting arguments of the matrix. The sequential solution consists of two nested loops, one each for the rows and columns of the matrix (Table 7-1). Each point in the matrix is evaluated in sequence, and the results accumulate in a result matrix.

Using the techniques described earlier, it can be seen that the program contains a nested loop where each element in the result is computed independently of all other elements. This problem easily converts to a parallel solution.

The inner loop in the sequential solution computes the values for one row of the matrix. In the parallel version, many rows can be computed simultaneously by different processes (Table 7-2). Using the data-decomposition model, a master program launches multiple worker processes with Linda's "eval" primitive (creating an active tuple) and distributes tasks to the workers with the "out" primitive.

Each task represents one row of the matrix. Each worker repeatedly retrieves a task tuple from the shared tuple space, computes the values for the corresponding row of the result matrix, and "outs" the result row back to the tuple space. The workers continue to retrieve tasks with the "inp" primitive until no tasks remain. Meanwhile, the master process is retrieving the result rows from the tuple space to assemble the complete result matrix.

Once a parallel solution is coded, it is often necessary to evaluate the performance of the program and modify the design for greater efficiency. It is useful to track the performance of a parallel program as processes are added.

Ideally, a linear speed-up should be seen as more processes are added, but in practice this improvement will not go on indefinitely. The programs typically show a near-linear speed-up for the first several processors, and then fall off

TABLE 7-1

FUNCTION MAPPING (SEQUENTIAL SOLUTION)

```
#define N 300

double results [N] [N];
extern double f(int, int, int);

void main () {
    int row, column;
    // loop through the rows
    for (row = 0; row < N; row + + ) {
        // and do each item in each row
        for (column = 0; column < N; column + + ) {
            results [row] [column] = f(row, column, N);
        }
    }
}
```

TABLE 7-2

```
FUNCTION MAPPING (PARALLEL SOLUTION)

#include <Linda.h>

#define N 300

double results [N] [N];
extern double f(int, int, int);

int worker () {
     int row, column;
     double line [N]

          while (inp("task", ?row)) {
               for (column + ); columns < N; column + + ){
                    line[column] = f(row, column, N);
               }
               out ("line", row, line:N);
          }
          return 0;
}
void Imain (int argc, char ** argv)
{
     int row, workers;

     /*put tasks into tuple space*/
     for (row =0; row < N; row + + ) {
          out("task",row);
     }

     /*create workers*/
     for (workers = 0; workers <atoi(argv [1]; workers + + ) {
          eval("worker", worker ());
     }

     /*remove results from tuple space*/
     for (row = 0; row < N; row + + ) {
          in("line", row, ?result[row] :N);
     }
     /*clean up workers*/
     for (row = 0; row < atoi(argv[1]); row + + ) {
          in("worker",0);
     }
}
```

sharply. The reason the fall-off takes place is that communication overhead begins to dominate the computation time. Each program will have its own particular point where communication and computation are optimally balanced, which can typically be found by experimentation.

The function-mapping example divided the problem into tasks where each task represented the computation of one row of the result matrix. It could have been constructed so that each task represented either one element of the result matrix or a block of rows. Linda offers the flexibility to try different options without major program modification. As a result, it is easier to arrive at the optimal parallel solution. Last, the program can be moved to other parallel machines to ensure that the investment in software development will not become obsolete as hardware improves.

REFERENCE

1. Translating Sequential Programs to Parallel, Dan Weston, Cogent Research Inc., Beaverton, OR. Electronic Design, March 8, 1990, P81, West Hasbrouck Heights, NJ.

7.4   ISIS

Networks of inexpensive UNIX™-based workstations offer great promise as computational engines for a wide range of coarsely parallel applications. Unfortunately, under contemporary versions of the UNIX operating system, effective utilization of multiple machines can be clumsy at the shell level, and impractical at a program level.

The ISIS [1] resource manager gathers collections of heterogenous, idle workstations into a processor pool onto which tasks are scheduled, and deals with dynamic events such as crashes, the need to release a workstation when its owner resumes active use, and the introduction of new types of machines and software services while the system is running. The solution is easy to use, has been ported to a wide variety of UNIX platforms, and offers multiple interfaces oriented toward different classes of users. One of these mimics a traditional batch queueing system, whereas others are more dynamic and suitable for use in explicitly distributed software systems that exploit program and data replication to increase performance or fault-tolerance.

The resource manager has been applied to problems in computer aided design, simulation, scientific computing, distributed software development, graphics, and many other areas. These uses include existing applications that must be executed without modification as well as new software that benefits by making explicit use of the resource manager at the program level.

As networks have become increasingly prevalent, more and more users are encountering problems that can benefit from being run on multiple machines in parallel. As one example, VLSI circuit simulation typically involves many trial runs of a circuit using different combinations of inputs. Here, a single program is run many times with different inputs. Such a problem is ideal for solution using a network of conventional high-speed workstations. Indeed, since the communication requirements of such an application are minimal, a closely coupled multiprocessor is not needed and the workstation approach will be considerably more efficient than batch style execution on a shared supercomputer. This trend has created a need for automatic network resource management tools.

Unfortunately, under contemporary versions of the UNIX operating system, effective utilization of collections of machines can be awkward. The available tools include programs such as ruptime, rwho, rsh, and rlogin. Using these, it is difficult to determine which machines are the most appropriate ones to use, and there is no network-wide scheduling mechanism to enforce fairness. There are no tools to check the status of active programs on the network, or to prioritize the use of certain machines in favor of certain sets of users. If an application may run for hours, days, or even weeks, and must be automatically monitored and restarted in the event of failure, there may be no human in the loop at the time a network scheduling activity is required. UNIX provides little help in these areas, forcing users to cobble together approximate solutions using mail to report completion status, periodically running ps to check on job activity, and so forth.

The resource manager solves these problems by providing an easy to use, portable, fault-tolerant mechanism for job management and monitoring in a distributed environment.

Much research has been done on the prospect of exploiting the power of distributed networks of workstations. However, most research provides only several pieces of the puzzle, not the whole solution. Shared memory models, reliable broadcast protocols, remote procedure calls, and so on, all contribute to the ability to distribute an application across the network. However, the development of the ISIS Distributed Toolkit technology addresses the broader, more complete picture.

Basically, ISIS is a subroutine package that employs protocols built over UDP to ensure that messages will be delivered reliably and in order; it adds headers to messages and delays messages on arrival (if necessary) to accomplish this. This reliable, consistent ordering of messages is called "Virtual Synchrony." Events such as multicast and detection of failures are atomic in a virtually synchronous setting; events appear to happen *one-at-a-time* and in a consistent order at all sites.

ISIS provides a rich set of distributed programming techniques based on these protocols. Central to ISIS is the notion of a *process group*. These groups are a lightweight programming construct: A single process can belong to arbitrarily many groups, and there is minimal overhead in being a member of a group. A process can dynamically join and leave groups, and groups can span multiple machines. ISIS provides a state transfer utility and several multicast and unicast communication primitives, with differing levels of ordering guarantees, for point-to-point and group comunication. A multicast can be directed to all members of a group, and zero or more will respond, depending on the needs of the particular application.

Although the overhead of using a package such as ISIS may negatively impact some applications such as real-time systems, its effect on an application such as the resource manager is negligible. In the resource manager, the time to process a job request is heavily biased by the fork/exec system calls. The communication and ordering overhead of ISIS represents only about 20 of the 450 ms average response time for a job request on a SUN sparcstation1.

All three pieces of the resource manager system are built on the ISIS Distributed Toolkit, and use causal and atomic multicase, process group communication, data replication, and failure detection. The use of ISIS relieved much of the difficulty associated with failures, synchronization, consistency, and network communications. Taken together with the wide range of tools represented in the toolkit, this approach led to major improvements in the robustness of the system. More information on ISIS, and performance data, can be found in the references [2–4].

The resource manager system consists of three parts: the resource manager server; the resource manager stubs; and the various user-interfaces. The resource manager server normally runs on two to five machines that are to be included in the resource pool. The user-interface programs run on end-users workstations (see Figure 7.4-1).

The resource manager server manages a customizable database of registered machines and services. Normally several copies of the resource manager server are instantiated, forming a process group with fully replicated data for fault-tolerance. All communication between the remote stubs, the resource manager server process group, and the user-interfaces use the ISIS reliable multicast protocols, cbcast, and abcast. Using the ordering guarantees of these protocols, every event is seen in the same order at each server, so if a server or stub crash, each remaining server has identical, consistent information on which to act.

Each resource manager server executes the same code in parallel and responds to incoming events by using the ISIS lightweight tasking subsystem to fork off tasks as event handlers. The event types recognized by the resource manager include the following.

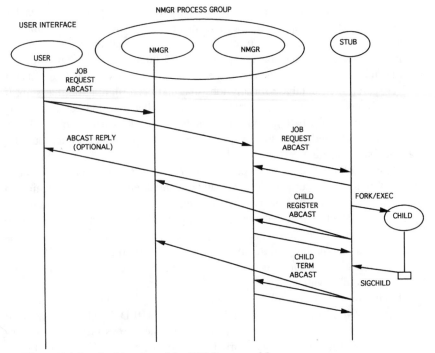

Figure 7.4-1.   Architecture of the ISIS Resource Manager.

- Resource manager process group join or leave event
- Resource manager stub join or leave event
- Resource manager stub load update message
- User-interface request message
- API request message
- Child (job) registration message
- Child (job) termination message

Each resource manager server is sensitive to server and stub join/leave events. On receiving a server join request, the oldest member of the server process group initiates a state transfer to the joining server. All events are briefly suspended while the transfer takes place, insuring that all database information is received by the joining member in a consistent manner. On a server leave or failure event, the surviving members of the server group reconfigure, and any stubs connected through the failed server reconnect to a surviving server.

The resource manager implements user-definable, priority-based queues that mimic traditional batch queueing mechanisms, but that provide the *fault-tolerant* execution of jobs. On a stub join event, the servers signal the batch queueing mechanism that a new machine has joined the resource pool, and search for any job that matches the new machine's specifications. On a stub leave event, the servers search their database for any jobs that had been running on that stub. If jobs are found, they attempt to restart them on another machine before deleting the failed stub's information from their database. The jobs from the failed stub will either be started on a different stub, or will be placed in a queue until a matching stub machine becomes available.

When a job request is received, a newly created task will add the job-related information to the resource manager's database and attempt to find a machine on which to execute the job. A pattern matching algorithm is run in combination with a simple but effective load-balancing algorithm to select a machine on which to run a submitted job request. If a matching stub machine is found, a message is sent to the stub containing all the information necessary to execute the job. If no matching machine is found, the job is placed in a batch queue until a matching machine becomes available.

Each stub is responsible for sending in a CHILD_REGISTER and CHILD_TERM message for each job it starts. The CHILD_REGISTER message contains the remote stub's machine name and the new job's process id and informs the server that the job was started successfully. The CHILD_TERM message returns the job's exit status and signals the normal or abnormal completion of the job. On receiving a CHILD_TERM message, the resource manager may signal the batch queueing mechanism that the stub is available to run another job. Depending on the job specifications and the exit status, the server may take other actions as well, such as automatically restarting the job or sending mail to the initiator of the job.

All machine- and job-related information is stored in a replicated database by each server and is available for status queries via one of the user-interface programs.

Following is an example of a resource manager database for a small network that specifies key words, machine types, individual machines, a batch queue, and an administrator's login id (for email notification upon failure events).

key	fpu	/* Has hardware fpu */
key	mathlab	/* Has mathlab license */
key	xfpu	/* Has high speed FPU support */
key	sparc	/* Sparc architecture */
key	mc68020	/* MC68020 architecture */
key	mips	/* mips architecture */
mtype	sun 3	= {mc68020,fpu, mem=8, rating=1.0}

```
mtype        sparc1+      = {sparc, mem=16, rating=8.0}
mtype        sparc2       = {sparc, fpu, xfpu, mem=16, rating=16.0}
mtype        mips         = {mips, fpu, xfpu, mem=16, rating=16.0}
mtype        hpux         = {hpux, fpu, xfpu, mem=16, rating=16.0}

machine flute             = sparc1+
machine viola             = {mips, mathlab,mounts={/usr/fsys/fs1, /usr/fsys/fs2}}
machine bongo             = {sparc1+,mem=32,mounts={/usr/fsys/fs1}}

add_q NEW_Q               = {nq_maxsize=120,nq_maxactive=
                            100,nq_maxrun=200,\
                            nq_maxmem=20}

admin_id tclark
```

The resource manager recognizes many options within job specifications for services such as the following.

- Sending mail on completion of a job
- Restricting the number of jobs simultaneously running on a machine
- Automatically restarting reliable server-based applications
- Scheduling cron jobs or sequentially related jobs
- Requesting specific machines for execution
- Copying files into and/or out of the execution directory
- Arbitrary, user-defined attributes used to match specifically equipped stub machines to job requests, for example, machines licensed to run specific software

The resource manager stub is run on each workstation that wishes to participate in the resource pool. When this program is instantiated, it first reads in its machine-specific information from an initialization file and then registers with the resource manager server, communicating its host name, machine type, current load, and mounted file systems. Once registered, its only activity is to send in its current load information at regular intervals and await incoming job requests from the server. In this state, the cpu overhead of running the stub on a workstation is insignificant.

When a job request is received by a stub, the message is unpacked, retrieving the binary name to be run, the arguments, and the environment variables. It will then use the UNIX fork/exec system calls to execute the requested binary. The fork/exec system calls represent the major portion of the system's performance, the actual communication of the job request taking approximately 20 ms. After the fork/exec, the stub sends in a CHILD_REGISTER message to the resource manager group, containing the process id of the spawned job. The UNIX wait() system call is used to reap terminated child processes and their associated exit status, and this information is passed on to the resource manager server(s) via a CHILD_TERM message.

The stub program is highly configurable. Machine-specific information is associated with the stub through the initialization file. This file contains attributes such as the machine type, the amount of physical memory, the "rating" of its processor (in user-definable units), the number of processors available (for multiprocessors), and other user-definable attributes such as software licenses or special hardware options (e.g., floating point processors).

The stub's command line arguments allow the specification of the following behavioral options.

- Go to sleep for a specified time interval if console login takes place
- Go to sleep if keyboard/mouse activity occurs on the host console
- Go to sleep if the host cpu load goes beyond a specified threshold level
- The ability to "plug-in" a user-defined load monitor routine
- The specification of the execution directory where job requests will be run (the default is /tmp/nmgr)

The resource manager has three types of user-interfaces: a shell-level command line interface; an X Windows/Motif graphical user-interface; and an applications programmer interface (API). The typical response time for a completed job request (via netexec or xnmgr) is on the order of half a second.

A shell-level interface has been implemented using the names netexec, netkill, netstatus, reserve, and unreserve—all links to the same binary. These interfaces allow the user to start, monitor, and kill jobs, as well as the ability to "reserve" machines for future use and "unreserve" them, as desired, in a transparent manner.

Example 1   net_exec "{binary={sparc=/usr/u/fred/test},min=3,mem>10,\
                       send_mail}"

This job request will start three copies of the binary /usr/u/fred/test on sparc-based workstations with physical memory greater than 10 MB, and will send email to the user on job completion.

Example 2   netexec "test1={binary={sparc=/usr/u/fred/sparc/test,\
                       mips=/usr/u/fred/mips/test},args={4},\
                       min=4, mounts=/usr/u/fred/data}"

This job specification will execute four copies of the binaries /usr/u/fred/ sparc/test or /usr/u/fred/mips/test with the argument "4" on sparc or mips-based workstations that have the /usr/u/fred/data file system mounted. The job will be run under the job name "test1", which can be used by the netkill program to kill the job(s) if so desired.

Example 3   netstatus -j
                  will output:

Job Name	Binary Name	Where
test1.tclark	/usr/u/fred/sparc/test	flute.cs.cornell.edu
test1.tclark	/usr/u/fred/mips/test	viola.cs.cornell.edu
test1.tclark	/usr/u/fred/sparc/test	1 instances queued
test1.tclark	/usr/u/fred/mips/test	1 instances queued

The graphical interface, based on the Motif widget set, is called xnmgr. This interface, using popup menus, push buttons, sliders, and text entry widgets, provides full functionality access to the resource manager, including the management of jobs (start, suspend, restart, kill), the creation of batch queues, and more detailed status information.

An Application Manager Interface API was added to the resource manager to provide the ability to communicate at the program level. The API provides an entry point for generic network manager commands, such as adding machine or service types or making job or status requests.

The evolution of computing networks has moved through a series of stages. In the 1980s, workstations were relatively new on the scene and represented a huge amount of computing power with respect to time-shared processors. Applications needing still more power generally turned to mainframes, special purpose mini-supercomputers and supercomputers. It is only recently that networks have become so common and workstations so fast that the idle processing power of a typical network routinely exceeds the processing capacity of midrange supercomputing systems. Indeed, within a few years, the aggregate capacity of a network of high-end workstations will outperform all but the fastest supercomputers. Utilities like the ISIS resource manager place this huge computational resource at the disposal of nonexpert UNIX programmers, permitting a style of cooperation and sharing what would previously have been difficult and impractical. Moreover, the resource manager can play a valuable role in applications that require high reliability and fault-tolerance, or wish to exploit parallelism for improved response time.

The resource manager is also interesting as an illustration of how new technology, such as the ISIS programming environment, can provide the necessary tools to allow the relatively easy development of a reliable, distributed application. Using ISIS permits one to focus on the functionality of the subsystem rather than the intricate and subtle problems raised by network protocols and fault-tolerance.

Looking to the future, it seems reasonable to predict that a new wave of high reliability, easily employed distributed programming tools will transform the UNIX networks of the 1990s into much more flexible and highly integrated distributed computing environments than are presently available.

REFERENCES

1. Using the ISIS Resource Manager for Distributed, Fault Tolerant Computing, Timothy Clark and Kenneth Birman, Dept. of Computer Science, Cornell University, June 23, 1992.

2. Birman, Kenneth P. and Joseph, Thomas A., Exploiting virtual synchrony in distributed systems. In *Proceedings of the Eleventh ACM Symposium on Operating Systems Principles*, pp. 123–138, Austin, Texas, November 1987. ACM SIGOPS.

3. Birman, Kenneth P. and Joseph, Thomas A., Reliable communication in the presence of failures. *ACM Transactions on Computer Systems*, Vol. 5, No. 1, pp. 47–76, February 1987.

4. Birman, Kenneth P., Joseph, Thomas A., Kane, Kenneth, and Schmuck, Frank, *ISIS—A Distributed Programming Environment User's Guide and Reference Manual*, 1st ed. Dept. of Computer Science, Cornell University, March 1988.

7.5 MESSAGE PASSING INTERFACE (MPI)

MPI stands for Message Passing Interface. The goal of the Message Passing Interface (MPI) effort, simply stated, is to develop a widely used standard for writing message-passing programs[1]. As such the interface should establish a practical, portable, efficient, and flexible standard for message passing [1].

Message passing is a paradigm used widely on certain classes of parallel machines, especially those with distributed memory. Although there are many variations, the basic concept of processes communicating through messages is well understood. Over the last ten years, substantial progress has been made in casting significant applications in this paradigm. Each vendor has implemented its own variant. More recently, several systems have demonstrated that a message passing system can be efficiently and portably implemented. It is thus an appropriate time to try to define both the syntax and semantics of a core of library routines that will

[1]Based on "MPI Frequently Asked Questions," compiled by Nathan Doss (doss@ERC.MsState.Edu), with permission.

be useful to a wide range of users and efficiently implementable on a wide range of computers.

In designing MPI the MPI Forum sought to make use of the most attractive features of a number of existing message passing systems, rather than selecting one of them and adopting it as the standard. Thus, MPI has been strongly influenced by work at the IBM T.J. Watson Research Center, Intel's NX/2, Express, nCUBE's Vertex, p4, and PARMACS. Other important contributions have come from Zipcode, Chimp, PVM, Chameleon, and PICL.

The main advantages of establishing a message-passing standard are portability and ease-of-use. In a distributed memory communication environment in which the higher level routines and/or abstractions are built on lower level message passing routines the benefits of standardization are particularly apparent. Furthermore, the definition of a message passing standard, such as that proposed in MPI, provides vendors with a clearly defined base set of routines that they can implement efficiently, or in some cases provide hardware support for, thereby enhancing scalability.

The Message Passing Interface Forum (MPIF), with participation from over 40 organizations, has been meeting since November 1992 to discuss and define a set of library interface standards for message passing. MPIF is not sanctioned or supported by any official standards organization.

The technical development was carried out by subgroups, whose work was reviewed by the full committee. During the period of development of the Message Passing Interface, many people served in positions of responsibility and are listed in the following.

- Jack Dongarra, David Walker; Conveners and Meeting Chairs
- Ewing Lusk, Bob Knighten; Minutes
- Marc Snir, William Gropp, Ewing Lusk; Point-to-Point Communications
- Al Geist, Marc Snir, Steve Otto; Collective Communications
- Steve Otto, Editor
- Rolf Hempel, Process Topologies
- Ewing Lusk, Language Binding
- William Gropp, Environmental Management
- Jams Cownie, Profiling
- Anthony Skjellum, Lyndon Clarke, Marc Snir, Richard Littlefield, Mark Sears; Groups, Contexts, and Communicators
- Steven Huss-Lederman, Initial Implementation Subset

The MPI standardization effort involved about 60 people from 40 organizations mainly from the United States and Europe. Most of the major vendors of concurrent computers were involved in MPI, along with researchers from universities, government laboratories, and industry. The standardization process began with the Workshop on Standards for Message Passing in a Distributed Memory Environment sponsored by the Center for Research on Parallel Computing, held April 29–30, 1992, in Williamsburg, VA. At this workshop the basic features essential to a standard message-passing interface were discussed, and a working group established to continue the standardization process.

A preliminary draft proposal, known as MPI-1, was put forward by Dongarra, Hempel, Hey, and Walker in November 1992, and a revised version was completed in February 1993. MPI-1 embodied the main features that were identified at the Williamsburg workshop as being necessary in a message passing standard. Since MPI-1 was primarily intended to promote discussion and "get the ball rolling," it focused mainly on point-to-point communications. MPI-1 brought to the forefront a number of important standardization issues, but did not include any collective communication routines and was not thread-safe.

In November 1992, a meeting of the MPI working group was held in Minneapolis, at which it was decided to place the standardization process on a more formal footing, and to generally adopt the procedures and organization of the High Performance Fortran Forum. Subcommittees were formed for the major component areas of the standard, and an email discussion service established for each. In addition, the goal of producing a draft MPI standard by the Fall of 1993 was set. To achieve this goal the MPI working group met every 6 weeks for 2 days throughout the first 9 months of 1993, and presented the draft MPI standard at the Supercomputing '93 Conference in November 1993. These meetings and the email discussion together constituted the MPI Forum, membership of which has been open to all members of the high performance computing community.

It was decided at the final MPI meeting (Feb. 1994) that plans for extending MPI should wait until people have had some experience with the current version of MPI. The MPI Forum held a session at Supercomputing '94 to discuss the possibility of an MPI-2 effort. It was decided that it was not time to begin official meetings for MPI-2.

In the discussion of possible MPI-2 extensions, held at the end of the February 1994 meeting, the following items were mentioned as possible areas of expansion.

- I/O
- Active messages
- Process startup
- Dynamic process control
- Remote store/access

- Fortran 90 and C++ language bindings
- Graphics
- Real-time support
- Other "enhancements"

Working together, IBM Research and NASA Ames have drafted MPI-IO, a proposal to address the portable parallel I/O problem. In a nutshell, this proposal is based on the idea that I/O can be modeled as message passing: Writing to a file is like sending a message, and reading from a file is like receiving a message. MPI-IO intends to leverage the relatively wide acceptance of the MPI interface in order to create a similar I/O interface.

The current proposal represents the result of extensive discussions (and arguments) but is by no means finished. Many changes can be expected as additional participants join the effort to define an interface for portable I/O.

Several free MPI implementations are currently available.

- Argonne National Laboratory/Mississippi State University implementation. Available by anonymous ftp at ftp://info.mcs.anl.gov/pub/mpi.
- Edinburgh Parallel Computing Centre CHIMP implementation. Available by anonymous ftp at ftp://ftp.epcc.ed.ac.uk/pub/chimp/release/chimp.tar.Z.
- Mississippi State University UNIFY implementation. The UNIFY system provides a subset of MPI within the PVM environment, without sacrificing the PVM calls already available. Available by anonymous ftp at ftp://ftp.erc.msstate.edu/unify.
- Ohio Supercomputer Center LAM implementation. A full MPI standard implementation for LAM, a UNIX cluster computing environment. Available by anonymous ftp at ftp://tbag.osc.edu/pub/lam.

The freely available versions of MPI port to a wide variety of platforms. New ports are a common occurrence. Please check with the authors of these implementations for the authoritative list of platforms supported.

- MPICH is supported on a variety of parallel computers and workstation networks. Parallel computers that are supported include: IBM SP1, SP2 (using various communication options); TMC CM-5; Intel Paragon, IPSC860, Touchstone Delta; Ncube2; Meiko CS-2; Kendall Square KSR-1 and KSR-2; SGI; and Sun Multiprocessors.

Workstations supported include: Sun4 family running SunOS or Solaris; Hewlett-Packard; DEC 3000 and Alpha; IBM RS/6000 family; SGI; and Intel 386- or 486-based PC clones running the LINUX or FreeBSD OS.

- CHIMP supports the following platforms.

    Sun workstations running SunOS 4.1.x or Solaris 2.x

    SGIs with IRIX 4 or IRIX 5

    IBM RS/6000 running AIX3.2

    Sequent Symmetry

- DEC Alpha AXP running OSF1

- Meiko Computing Surface with transputer, i860 or SPARC nodes

- UNIFY runs atop PVM and therefore is portable to the same platforms as PVM.

- LAM is portable to most unix systems and includes standard support for the following platforms.

    Sun 4 (sparc), SunOS 4.1.3

    Sun 4 (sparc), Solaris 2.3

    SGI IRIX 4.0.5

    IBM RS/6000, AIX v3r2

    DEC AXP, OSF/1 V2.0

    HP 9000/755, HP-UX 9.01

The official postscript version of the MPI document can be obtained from netlib at ORNL by sending a mail message to netlib@ornl.gov with the message "send mpi-report.ps from mpi".

Argonne National Lab also provides a hypertext version available through the WWW at http://www.msc.anl.gov/mpi/mpi-report/mpi-report.html.

The following is a list of URLs that contain MPI related information.

- Netlib Repository at UTK/ORNL
  (http://www.netlib.org/mpi/index.html)

- Argonne National Lab
  (http://www.mcs.anl.gov/mpi)

- Mississippi State University, Engineering Research Center
  (http://www. erc.msstate.edu/mpi)

- Ohio Supercomputer Center, LAM Project
  (http://www.osc.edu/larn.html)

- Australian National University
  (file://dcssoft.anu.edu.au/pub/www/dcs/ cap/mpi/mpi.html)

A bibliography (in BibTeX format) of MPI related papers is available by anonymous ftp at ftp://ftp.erc.msstate.edu/pub/mpi/bib/MPI.bib. It is also available on the WWW from http://www.erc.msstate.edu/mpi/mpi-bib.html. Additions and corrections should be sent to doss@ERC.MsState.Edu.

MPI-related books are available from William Gropp, Ewing Lusk, and Anthony Skjellum, *USING MPI: Portable Parallel Programming with the Message-Passing Interface* (MIT Press, 1994, 328 pages, paperback, $24.95).

Ian Foster's online book entitled, *Designing and Building Parallel Programs* (ISBN 0-201-57594-9; published by Addison-Wesley) includes a chapter on MPI. It provides a succinct and readable introduction to an MPI subset.

The MPI standard has been published as a journal article in the *International Journal of Supercomputing Applications* (Vol. 8, No. 3/4, 1994).

A copy of the ascii version of the MPI FAQ is posted monthly to comp.parallel.mpi; news.answers; and comp.answers. The ascii version can also be obtained through anonymous ftp from ftp://rtfm.mit.edu/pub/usenet/news.answers/mpi-faq or those without FTP access can send e-mail to mail-server@rtfm.mit.edu with "send usenet/news.answers/mpi-faq" in the message body.

The MPI FAQ was compiled by Nathan Doss (doss@ERC.MsState.Edu), with assistance and comments from others. MPI Frequently Asked Questions © 1994 by Mississippi State University. It may be reproduced and distributed in whole or in part, subject to the following conditions.

- This copyright and permission notice must be retained on all complete or partial copies.

- Any translation or derivative work must be approved by me before distribution.

- If you distribute MPI Frequently Asked Questions in part, instructions for obtaining the complete version of this manual must be included, and a means for obtaining a complete version free or at cost price provided.

REFERENCE

1. Smir, Marc, Otto, Steve W., Huss-Lederman, Steven, Walker, David W., Dongurra, Jack, *MPI: The Complete Reference,* MIT Press, Cambridge, MA, 1996.

Chapter 8

Software

Software is the hard part of parallel supercomputing. It is not as hard as many people perceive it to be, but it is harder than conventional programming. Mainly, people perceive it to be so hard because it is different from programming sequential machines and many users try to parallel program without taking any training specifically directed at developing parallel applications.

The three sections presented here are a mixed bag; Compositional C++ describes a parallel object oriented programming language; High Performance Fortran is a mainstream parallel version of Fortran developed by a consortium of all the major vendors and universities in the parallel supercomputing community; Paragraph is a free software package from Oak Ridge National Laboratory that facilitates applications development by providing the developer with diverse visual diagnostics for tuning the performance of parallel applications.

8.1   COMPOSITIONAL C++

Programming languages provide mechanisms for structuring programs as a composition of component units. Properties of programs composed using sequential or functional composition can be derived from properties of their components.

Sequential composition and functional composition follow the principle of implementation hiding: Programmers can compose program specifications without being concerned with program implementation.

In parallel programs, however, implementation hiding is unsatisfactory. To prove properties of a program composed in parallel we must (in general) give proofs for the components such that one component does not interfere with the proof for the other. To prove properties of the composed program we have to use details of the components that we would rather hide.

One way of hiding implementation is to design parallel composition such that properties of components are also properties of the composed program. In this case we need not concern ourselves with implementations of the components, nor need we be concerned about one component interfering with the proof of another. The key issue, then, is to design parallel composition so that the whole inherits properties of the parts.

The rules for proper parallel composition are designed to yield the inheritance property. The central idea in obtaining compositionality is that a process has private variables that cannot be referenced by other processes, and shared variables that can be assigned values at most once. (A message sequence can be thought of as a list in which each element—a message—is assigned a value when the message is sent.) Examples of the use of single assignment variables can be found in the references [1].

One issue is what programming notation should be used to express proper parallel composition in a program [2]. A significant advantage of using small extensions to widely used sequential languages is that programmers can learn the extensions quickly, and use tools developed for the base language. This is the approach taken here: using C++ as the base language [3], because of its widespread use, and its support for abstract data types, object-oriented programming, and the specification of generic algorithms through templates and inheritance.

C++ can be augmented to create a parallel programming language; the resulting language is called Compositional C++, or CC++ [4]. The following discussion will focus on the essential aspects of CC++; a complete description of CC++ can be found in the references [5, 6]. CC++ is a flexible, concurrent object-oriented programming language particularly well suited to the implementation of large-scale parallel programs and high performance parallel programming libraries utilizing a range of parallel programming paradigms.

8.1.1   Determinism and Nondeterminism

Programmers developing scientific applications want their programs to be deterministic: Executions of the same program with the same inputs should produce the same outputs. Numerical analysts craft careful sequences of steps to avoid numerical instability, and they need to be guaranteed that the same sequence of steps will be executed in each execution. By contrast, reactive systems are nondeterministic because they deal with an uncertain environment.

Programmers can guarantee that CC++ programs are deterministic by following certain simple conventions—if they follow these conventions, then they have no proof obligation in demonstrating determinism. Programmers can, if they wish, use nondeterministic constructs. The ability to choose between deterministic and nondeterministic constructs, and to compose deterministic and nondeterministic programs, allows programmers to develop a variety of applications, including reactive systems with components that are scientific applications.

8.1.2   Paradigm Integration

No "best" parallel programming paradigm exists. Semaphores, monitors, message passing, and so on, all have their uses. Different paradigms can be appropriate even with a single program. CC++ was designed to facilitate the use of different programming paradigms, such as the following.

- Task, Data, and Object Parallel. Distributed arrays and distributed grids can be defined as classes in CC++, allowing data-parallel computations on these objects. In addition, CC++ supports cooperative processing and task parallelism. Furthermore, since CC++ is object-oriented, it supports the object-oriented paradigm.

- Declarative and Imperative. The extensions to C++ allow declarative programs to be written in CC++. However, programmers can continue to use the familiar imperative style of C++.

- Shared Variables and Message-Passing. Although processes in CC++ share variables, libraries of message-passing channels are provided so that the message-passing paradigm can be used if desired. Libraries of semaphores and monitors allow programmers to use the styles of their choice.

As its name implies, CC++ is based on C++ [7]. C++ is itself is an extension of the C programming language. C++ adds strict typing, function overloading, encapsulation, abstract data types, and object-oriented programing to C. These features make C++ a good language for implementing program libraries and large scale software systems.

A central concept in C++ is the *class*, which combines code and data into a single unit, or object. The data components of a class are called *data members*. The functions associated with a class are called *member functions*; in other object-oriented languages, member functions are often called *methods*. The subset of the data members and member functions that define the interface to the class are declared *public* and are accessible from outside the class. All other members are private and they are accessible only to member functions of the class. A class, therefore, forms a unit of encapsulation.

The advantages of an object-oriented approach to the design of parallel systems has long been recognized. Object-oriented systems provide well-defined interfaces, collocation of function and data, encapsulation of data, and data abstraction. These features encourage the construction of scalable parallel systems.

Consequently, the encapsulation and object-oriented features of C++ have made it the basis of a number of parallel programming systems.

CC++ is a pure superset of C++; it consists of C++ plus seven extensions. These extensions impact the language in the following areas.

- Flow of control
- Synchronization and communication
- Nondeterminism
- Locality and heterogeneity

CC++ is intended to be a general purpose parallel programming language. As such, parallelism in a CC++ program is explicit. Three constructs are available for parallel composition in CC++: the par block, the parfor statement, and the spawn statement.

8.1.3   The Par Block

The par block is the most basic means of specifying parallel composition in CC++. Its syntax is that of the compound statement in C++ with the keyword par preceding the block. An example of a par block is found in Figure 8.1-1. A par block can lexically contain any CC++ statement with the exception of the return statement and variable declarations. (A goto statement is allowed in a par block, but its use is restricted.) As seen in Figure 8.1-1, the statements in the block can be sequential blocks and par blocks can be nested.

A new thread of control is created for each top level statement in a par block; a par block terminates when all its statements terminate. However, there is no requirement that a par block terminate. Parallel execution of the statements in the par block is defined by fair, interleaved execution of the top level statements in

```
par {
    procedure1();
    { procedure2();
      par {
          procedure3();
          procedure4();
      }
      procedure5();
    }
    procedure6();
}
```

Figure 8.1-1.   An example of a par block.

the block; fairness means that every executable statement in a par block will execute eventually, even if the par block does not terminate.

With the exception of atomic functions the granularity at which the interleaving occurs is not defined. As an example, consider the par of Figure 8.1-2. Assume that all the statements in the par block terminate. Possible execution orderings of the par block include (but are not limited to) the following.

```
a1  a2  a3  b1  b2  b3
a1  b1  b2  b3  a2  a3
b1  a1  a2  b2  a3  b3
```

The sequential ordering of the function calls within statements $\underline{S}1$ and $\underline{S}2$ is maintained. However, even though these statements are sequential, their execution is interleaved.

The ability to create a collection of threads is not too useful unless threads can interact with each other.

The primary communication mechanism in CC++ is shared variables. This is not to say, however, that CC++ can only execute efficiently on a shared memory computer. The sharing is constrained in such a way as to make efficient execution on a range of parallel architectures possible.

8.1.4   The Parfor Statement

A par is useful when one needs to create a fixed number of threads, and that number is shown at compile time. Although recursion within a par block can be used to create an arbitrary number of concurrent threads, iteration is often a more natural and convenient means of expressing such computations. For this reason, CC++ has a parallel loop construct, the parfor.

The syntax of a parfor statement is exactly like the for statement in C++. The statement specifies loop initialization, a termination test, an index update expression, and the loop body. The loop body can be a simple statement, a sequential block, or a parblock. Note that in C++, the index variable of a loop is not constrained to be an integer and the termination test and index update expression can be any valid C++ expression. An example of a parfor statement is shown in Figure 8.1-3. Notice the use of the C++ feature that allows the declaration of an index variable to be placed in the parfor statement itself.

```
par {
    { a1(); a2(); a3(); } // Statement S1
    { b1(); b2(); b3(); } // Statement S2
}
```

Figure 8.1-2.   Parallel execution of two sequential blocks.

```
parfor (int index = 0 ; index < N ; index++) {
    a1(index) ;
    a2(index+1) ;
    a3(index) ;
}
```

Figure 8.1-3.   A parfor statement.

Each iteration of a parfor creates a new thread that executes in parallel with all other iteration bodies (and the rest of the computation as well). The threads have the same interleaved execution semantics of the par block. When all the loop bodies terminate, the parfor statement terminates. Variables declared in the initialization part of a parfor loop receive special treatment. Within each thread, a local copy of the each index variable is created and initialized with the values of the index variables at the time of thread creation. Thus in Figure 8.1-3, each loop body will have a local variable called index and its value will be set to the value of index in the parfor loop at the time the thread for the loop body was created. Regardless of the execution order of the loop bodies, the correct value of index will be available in the loop body.

8.1.5   The Spawn Statement

The third and final method for specifying concurring execution in CC++ is the spawn statement. The termination criteria associated with par blocks and parfor statements imposes structure on a concurrent computation. When the statement terminates, all parallel computation is complete and any postconditions associated with the block hold. However, there are situations in which this structure is a hindrance rather than a help. An example is a program that sets up a network of interconnected servers. In this case, one would like to start a new thread of execution for each server and have the program proceed, independent of the termination of the thread. Although this could be done using a parfor or par block, one has to go to some effort to work around the fact that the servers must terminate before the statement after the par block or parfor statement executions. The situation would be simplified if one could start a new thread of control and proceed to the next statement immediately. This is exactly what the spawn statement does. An example of a spawn statement is found in Figure 8.1-4.

A spawn statement executes an arbitrary CC++ expression in a new thread of control. The execution of the statement is interleaved fairly with the rest of the

```
spawn x + y +g(z) ;
```

Figure 8.1-4.   A spawn statement.

program. Unlike the par and parfor, a spawn statement terminates immediately, regardless of the status of the process that was spawned.

8.1.6   Comparison with Other Approaches

Parallel composition as defined by par blocks can be found in a number of other parallel programming notations. For example, a par is equivalent to the use of cobegin and coend in reference [8] and the parallel composition operation in PCN.

The use of par blocks differs from most other concurrent object-oriented languages in that with a par block, multiple threads of control exist within a single object. Other languages tend to associate thread creation with object creation [9, 10]. Consequently, only one thread of control is ever associated with an object. The spawn statement in CC++ can be used to achieve the same effect.

The advantage of par blocks and parfor statements over the tying thread creation to object creation approach is twofold. First, these statements are block-oriented, and they make parallelism within a block explicit; there is no question as to which statements execute in parallel and which in sequence. The second advantage is that one can associate a post condition with a par block or a parfor statement. When the statement terminates, the post condition can be asserted. This simplifies the process of reasoning about the behavior of the program.

Reasoning about the behavior of a program containing a spawn statement is more difficult. Because there is no way of knowing when the thread started by the spawn starts or completes, assertions about the spawn statement must state that a condition will hold at some unknown point in the future. Thus one has no choice but to resort to a temporal operator such as the leads-to operator.

REFERENCES

1. Foster, Ian and Taylor, Stephen, *STRAND: New Concepts in Parallel Programming,* Englewood Cliffs, NJ, Prentice-Hall, 1989.

2. Kuck, David, Gajski, Daniel, et al. A second opinion on data flow machines and languages, *Computer,* Vol. 15, No. 2, pp. 58–69, February 1982.

3. Ellis, Margaret A. and Stroustrup, Bjarne, *The Annotated C++ Reference Manual,* New York, Addison-Wesley, 1990.

4. Chandy, K. Mani and Kesselman, Carl, Compositional C++: Compositional Parallel Programming, Jet Propulsion Laboratory, Pasadena, CA; latest copies of CC++ papers are available by anonymous ftp at csvax.cs.caltech.edu under comp/docs.

5. Chandy, K. Mani and Kesselman, Carl, The CC++ language definition. Technical Report Caltech-CS-TR-92-02, California Institute of Technology, Pasadena, CA, 1992.

6. Chandy, K. Mani and Kesselman, Carl, CC++: A declarative concurrent object-oriented programming notation. Technical Report Caltech-CS-TR-92-01, California Institute of Technology, 1992.

7. Stroustrup, Bjarne, *The C++ Programming Language,* 2nd ed. New York, Addison-Wesley, 1991.

8. Owicki S. and Gries, D., An axiomatic proof technique for parallel programs I, *Acta Informatica,* Vol. 6, No. 1, pp. 319–340, 1976.

9. Bershad, Brian, Lazowska, Edward, and Levy, Henry, Presto: A system of object-oriented parallel programming. *Software: Practice and Experience,* Vol. 18, No. 8, pp. 713–732, August 1988.

10. Grimshaw, Andrew S., An introduction to parallel object-oriented programming with Mentat. Computer Science Report TR-91-07, University of Virginia, Charlottesville, VA, 1991.

8.2   HIGH PERFORMANCE FORTRAN

High Performance Fortran (HPF) language [1] is designed as a set of extensions and modifications to the established International Standard for Fortran (ISO/IEC 1539:1991(E) and ANSI X3.198-1992), informally referred to as "Fortran 90."

8.2.1   Goals and Scope
of High Performance Fortran

The goals of HPF were to define language extensions and feature selection for Fortran supporting the following.

*   Data parallel programming (defined as single-threaded, global name space, and loosely synchronous parallel computation)
*   Top performance on MIMD and SIMD computers with nonuniform memory access costs (while not impeding performance on other machines)
*   Code tuning for various architectures

The FORALL construct and several new intrinsic functions were designed primarily to meet the first goal, whereas the data distribution features and some other directives are targeted toward the second goal. Extrinsic procedures allow access to low-level programming in support of the third goal, although performance tuning using the other features is also possible.

HPF is an extension of Fortran 90. The array calculation and dynamic storage allocation features of Fortran 90 make it a natural base for HPF. The HPF language features fall into four categories with respect to Fortran 90.

1. New directives

2. New language syntax

3. Library routines

4. Language restrictions

The new directives are structured comments that suggest implementation strategies or assert facts about a program to the compiler. They may affect the efficiency of the computation performed, but do not change the value computed by the program. The form of the HPF directives has been chosen so that a future Fortran standard may choose to include these features as full statements in the language by deleting the initial comment header.

A few new language features, namely the FORALL statement and certain intrinsic functions, are also defined. They were made first-class language constructs rather than comments because they can affect the interpretation of a program, for example, by returning a value used in an expression. These are proposed as direct extensions to the Fortran 90 syntax and interpretation.

The HPF library of computational functions defines a standard interface to routines that have proven valuable for high performance computing including additional reduction functions, combining scatter functions, prefix and suffix functions, and sorting functions.

Full support of Fortran sequence and storage association is not compatible with the data distribution features of HPF. Some restrictions on the use of sequence and storage association are defined. These restrictions may in turn require insertion of HPF directives into standard Fortran 90 programs in order to preserve correct semantics.

8.2.2   New Features in High Performance Fortran

HPF extends Fortran 90 in several areas, including the following.

- Data distribution features
- Parallel statements
- Extended intrinsic functions and standard library
- EXTRINSIC procedures
- Parallel I/O statements
- Changes in sequence and storage association

Modern parallel and sequential architectures attain their fastest speed when the data accessed exhibit locality of reference. Often, the sequential storage order

implied by FORTRAN 77 and Fortran 90 conflicts with the locality demanded by
the architecture. To avoid this, HPF includes features that describe the collocation
of data (ALIGN) and the partitioning of data among memory regions (DISTRIB-
UTE). Compilers may interpret these annotations to improve storage allocation
for data, subject to the constraint that semantically every data object has only a
single value at any point in the program.

To express parallel computation explicitly, HPF offers a new statement and a
new directive. The FORALL construct expresses assignments to sections of arrays; it is similar in many ways to the array assignment of Fortran 90, but allows
more general sections and computations to be specified. The INDEPENDENT directive asserts that the statements in a particular section of code do not exhibit
any sequentializing dependences; when properly used, it does not change the
semantics of the construct, but may provide more information to the language
processor to allow optimizations.

Because HPF is designed as a high-level, machine-independent language,
there are certain operations that are difficult or impossible to express directly. For
example, many applications benefit from finely tuned systolic communications
on certain machines; HPF's global address space does not express this well. Extrinsic procedures define an explicit interface to procedures written in other paradigms, such as explicit message-passing subroutine libraries.

A goal of HPF was to maintain compatibility with Fortran 90. Full support of
Fortran sequence and storage association, however, is not compatible with the
goal of high performance through distribution of data in HPF. Some forms of as-
sociating subprogram dummy arguments with actual values make assumptions
about the sequence of values in physical memory that may be incompatible with
data distribution. Certain forms of EQUIVALENCE statements are recognized as
requiring a modified storage association paradigm. In both cases, HPF provides a
directive to assert that full sequence and storage association for affected variables
must be maintained. In the absence of such explicit directives, reliance on the
properties of association is not allowed. An optimizing compiler may then choose
to distribute any variables across processor memories in order to improve performance. To protect program correctness, a given implementation should provide a mechanism to ensure that all such default optimization decisions are consistent across an entire program.

The facilities for array computations in Fortran 90 will make it the programming language of choice for scientific and engineering numerical calculations on
high performance computers. Indeed, some of these facilities are already supported in compilers from a number of vendors. The introductory overview in the
Fortran standard states:

> Operations for processing whole arrays and subarrays (array sections) are included in
> [Fortran 90] for two principal reasons: (1) these features provide a more concise and
> higher level language that will allow programmers more quickly and reliably to de-

velop and maintain scientific/engineering applications, and (2) these features can significantly facilitate optimization of array operations on many computer architectures.

Other features of Fortran 90 that improve on the features provided in FORTRAN 77 include the following.

- Additional storage classes of objects. The new storage classes such as allocatable, automatic, and assumed-shape objects as well as the pointer facility of Fortran 90 add significantly to those of FORTRAN 77 and should reduce the use of FORTRAN 77 constructs that can lead to less than full computational speed on high performance computers, such as EDQUIVALENCE between array objects, COMMON definitions with nonidentical array definitions across subprograms, and array reshaping transformations between actual and dummy arguments.

- Modules. The module facilities of Fortran 90 enable the practice of design implementation using data abstractions. These facilities support the specification of modules, including user-defined data types and structures, defined operators on those types, and generic procedures for implementing common algorithms to be used on a variety of data structures. In addition to modules, the definition of interface blocks enables the application programmer to specify subprogram interfaces explicitly, allowing a high quality compiler to use the information specified to provide better checking and optimization at the interface to other subprograms.

- Additional intrinsic procedures. Fortran 90 includes the definition of a large number of new intrinsic procedures. Many of these support mathematical operations on arrays, including the construction and transformation of arrays. Also, there are numerical accuracy intrinsic procedures designed to support numerical programming, and bit manipulation intrinsic procedures derived from MIL-STD-1753.

HPF conforms to Fortran 90 except for additional restrictions placed on the use of storage and sequence association. Because the effort involved in producing a full Fortran 90 compiler, HPF is defined at two levels: subset HPF and full HPF. Subset HPF is a subset of Fortran 90 with a subset of the HPF extensions. HPF is full Fortran 90 with all of the approved HPF language features.

An important goal of HPF is to achieve code portability across a variety of parallel machines. This requires not only that HPF programs compile on all target machines, but also that a highly efficient HPF program on one parallel machine be able to achieve reasonably high efficiency on another parallel machine with a comparable number of processors. Otherwise, the effort spent by a programmer to achieve high performance on one machine would be wasted when the HPF code is ported to another machine. Although SIMD processor arrays, MIMD shared-memory machines, and MIMD distributed-memory machines use very different low-level primitives, there is sufficient broad similarity with respect to the fundamental factors that affect the performance of parallel programs on these machines. Thus, achieving high efficiency across different parallel machines with the same high level HPF program is a feasible goal. Although describing a full execution model is beyond the scope of this language specification, we focus here on two fundamental factors and show how HPF relates to them.

- The parallelism inherent in a computation
- The communication inherent in a computation

The quantitative cost associated with each of these factors is machine-dependent; vendors are strongly encouraged to publish estimates of these costs in their system documentation. Note that, like any execution model, these may not reflect all of the factors relevant to performance on a particular architecture.

The parallelism in a computation can be expressed in HPF by the following "parallel" constructs.

- Fortran 90 array expressions and assignment (including conditional assignment in the WHERE statement)
- Array intrinsics, including both the Fortran 90 intrinsics and the new intrinsic functions
- The FORALL statement and construct
- The INDEPENDENT assertion on DO loops
- The extrinsic procedure mechanism

The preceding features allow the explicit user specification of a high degree of potential parallelism in a machine-independent fashion. In addition, extrinsic procedures provide an escape mechanism in HPF to allow the use of efficient machine-specific primitives by explicitly executing on a set number of processors.

A compiler may choose not to exploit information about parallelism, for example, because of lack of resources or excessive overhead. In addition, some compilers may detect parallelism in sequential code by use of dependence analysis. This document does not discuss such techniques.

The interprocessor data communication that occurs during the execution of an HPF program is partially determined by the HPF data distribution directives. The compiler will determine the actual mapping of data objects to the physical machine and will be guided in this by the directives. The actual mapping and the computation specified by the program determine the needed actual communication, and the compiler will generate the code required to perform it. In general, if two data references in an expression or assignment are mapped to different processors, then communication is required to bring them together. The examples that follow illustrate how this may occur.

Clearly, there is a tradeoff between parallelism and communication. If all the data are mapped to one processor, then a sequential computation with no communication is possible, although the memory of one processor may not suffice to store all the program's data. Alternatively, mapping data to multiple processors may permit computational parallelism but also may introduce communications overhead. The optimal resolution of such conflicts is very dependent on the architecture and underlying system software.

The following examples illustrate the communication requirements of scalar assignment statements. The purpose is to illustrate the implications of data distribution specifications on communication requirements for parallel execution and does not necessarily reflect the actual compilation process.

Consider the statements that follow.

```
        REAL a(1000), b(1000), c(1000), x(500), y(0:501)
        INTEGER inx(1000)
!HPF$ PROCESSORS procs(10)
!HPF$ DISTRIBUTE (BLOCK) ONTO procs :: a, b, inx
!HPF$ DISTRIBUTE (CYCLIC) ONTO procs :: c
!HPF$ ALIGN x(i) WITH y(i+1)
        . . .
        a(i) = b(i)                          ! Assignment 1
        x(i) = y(i+1)                        ! Assignment 2
        a(i) = c(i)                          ! Assignment 3
        a(i) = a(i - 1) + a(i) + a(i + 1)    ! Assignment 4
        c(i) = c(i - 1) + c(i) + c(i + 1)    ! Assignment 5
        x(i) = y(i)                          ! Assignment 6
        a(i) = a(inx(i)) + b(inx(i))         ! Assignment 7
```

In this example, the PROCESSORS directive specifies a linear arrangement of 10 processors. The DISTRIBUTE directives recommend to the compiler that the arrays a, b, and inx should be distributed among the 10 processors with blocks of 100 contiguous elements per processor. The array c is to be cyclically distributed among the processors with $c(1)$, $c(11)$, . . . , $c(991)$ mapped onto processor procs(1); $c(2)$, $c(12)$, . . . , $c(992)$ mapped onto processor procs(2); and so on. The complete mapping of arrays x and y onto the processors is not specified, but their relative alignment is indicated by the ALIGN directive. The ALIGN statement causes $x(i)$ and $y(i + 1)$ to be stored on the same processor for all values of i, regardless of the actual distribution chosen by the compiler for x and y ($y(0)$ and $y(1)$ are not aligned with any element of x).

In Assignment 1 ($a(i) = b(i)$), the identical distribution of a and b ensures that for all i, $a(i)$ and $b(i)$ are mapped to the same processor. Therefore, the statement requires no communication.

In Assignment 2 ($x(i) = y(i + 1)$), there is no inherent communication. In this case, the relative alignment of the two arrays matches the assignment statement for any actual distribution of the arrays.

Assignment 3 ($a(i) = c(i)$) looks very similar to the first assignment; the communication requirements are very different because of the different distributions of a and c. Array elements $a(i)$ and $c(i)$ are mapped to the same processor for only 10% of the possible values of i. The elements are located on the same processor if and only if $[(i - 1)/100] = i - 1)$ mod 10. For example, the assignment involves no inherent communication (i.e., both $a(i)$ and $c(i)$ are on the same processor) if $i = 1$ or $i = 102$, but does require communication if $i = 2$.

In Assignment 4 ($a(i) = a(i - 1) + a(i) + a(i + 1)$), the references to array a are all on the same processor for about 98% of the possible values of i. The exceptions to this are $i = 100k$ for any $k = 1,2, \ldots ,9$, (when $a(i)$ and $a(i - 1)$ are on procs(k) and $a(i + 1)$ is on procs(k + 1)) and $i = 100k + 1$ for any $k = 1,2, \ldots ,9$ (when $a(i)$ and $a(i + 1)$ are on procs(k + 1) and $a(i - 1)$ is on procs(k)). Thus, except for "boundary" elements on each processor, this statement requires no inherent communication.

Assignment 5, $c(i) = c(i - 1) + c(i) + c(i + 1)$, although superficially similar to the last, has very different communication behavior. Because the distribution of c is CYCLIC rather than BLOCK, the three references $c(i)$, $c(i - 1)$, and $c(i + 1)$ are mapped to three distinct processors for any value of i. Therefore, this statement requires communication for at least two of the right-hand side references, regardless of the implementation strategy.

The final two assignments have very limited information regarding the communication requirements. In Assignment 6 ($x(i) = y(i)$) the only information available is that $x(i)$ and $y(i - 1)$ are on the same processor; this has no logical consequences for the relationship between $x(i)$ and $y(i)$. Thus, nothing can be said regarding communication in the statement without further information. In Assignment 7 ($a(i) = a(inx(i)) + b(inx(i)))$, it can be proved that $a(inx(i))$ and $b(inx(i))$ are always mapped to the same processor. Similarly, it is easy to deduce that $a(i)$ and $inx(i)$ are mapped together. Without knowledge of the values stored in inx, however, the relation between these two pairs of references is unknown.

```
a(:) = b(:)


! Assignment 2 (equivalent to Forall 2)
      a(1:1000) = c(1:1000)


! Assignment 3 (equivalent to Forall 3)
a(2:999) = a(1:998) + a(2:999) + a(3:1000)


! Assignment 4 (equivalent to Forall 4)
c(2:999) = c(1:998) + c(2:999) + c(3:1000)
```

Some array intrinsics have inherent communication costs as well. For example, consider the following.

```
         REAL a(1000), b(1000), scalar
!HPF$ PROCESSORS procs(10)
!HPF$ DISTRIBUTE (BLOCK) ONTO procs :: a, b
         ...
         ! Intrinsic 1
         scalar = SUM( a )


         ! Intrinsic 2
         a = SPREAD( b(1), DIM=1, NCOPIES=1000 )


         ! Intrinsic 3
         a = CSHIFT(a,-1) + a + CSHIFT(a,1)
```

8.2.3   Data Alignment and Distribution Directives

HPF adds directives to Fortran 90 to allow the user to advise the compiler on the allocation of data objects to processor memories. The model is that there is a two-level mapping of data objects to abstract processors. Data objects (typically array elements) are first *aligned* relative to one another; this group of arrays is then *distributed* onto a rectilinear arrangement of abstract processors. (The implementation then uses the same number, or perhaps some smaller number, of physical processors to implement these abstract processors. This mapping of abstract processors to physical processors is language-processor-dependent.)

Figure 8.2-1 illustrates the model.

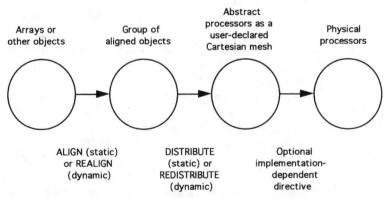

| Arrays or other objects | Group of aligned objects | Abstract processors as a user-declared Cartesian mesh | Physical processors |

| ALIGN (static) or REALIGN (dynamic) | DISTRIBUTE (static) or REDISTRIBUTE (dynamic) | Optional implementation-dependent directive | |

Figure 8.2-1.   Two-level mapping of data objects to abstract processors.

The underlying assumptions are that an operation on two or more data objects is likely to be carried out much faster if they all reside in the same processor, and that it may be possible to carry out many such operations concurrently if they can be performed on different processors.

Fortran 90 provides a number of features, notably array syntax, that make it easy for a compiler to determine that many operations may be carried out concurrently. The HPF directives provide a way to inform the compiler of the recommendation that certain data objects should reside in the same processor: If two data objects are mapped (via the two-level mapping of alignment and distribution) to the same abstract processor, it is a strong recommendation to the implementation that they ought to reside in the same physical processor. There is also a provision for recommending that a data object be stored in multiple locations, which may complicate any updating of the object but makes it faster for multiple processors to read the object.

There is a clear separation between directives that serve as specification statements and directives that serve as executable statements (in the sense of the Fortran standards). Specification statements are carried out on entry to a program unit, pretty much as if all at once; only then are executable statements carried out. (Although it is often convenient to think of specification statements as being handled at compile time, some of them contain specification expressions that are permitted to depend on run-time quantities such as dummy arguments; therefore, the values of these expressions may not be available until run time, specifically the very moment that program control enters the scoping unit.)

The general idea is that every array (indeed, every object) is created with *some* distribution onto *some* arrangement of processors. If the specification statements contain explicit B, then the distribution of A will be dictated by the distribution of B; otherwise, the distribution of A itself be may be specified explicitly. In either case, any such explicit declarative information is used when the array is created. (This model gives a better picture of the actual amount of work that needs to be done than a model that says, "The array is created in some default location, and then realigned and/or redistributed if there is an explicit directive." Using ALIGN and DISTRIBUTE specification directives does not have to cause any more work at run time than using the implementation defaults.)

In the case of an allocatable object, we say that the object is created whenever it is allocated. Specification directives for allocatable objects (and allocated pointer targets) may appear in the specification-part of a program unit, but take effect each time the array is created, rather than on entry to the scoping unit.

If an object A is aligned (statically or dynamically) with an object B, which in turn is already aligned to an object C, this is regarded as an alignment of A with C directly, with B serving only as an intermediary at the time of specification. (This matters only in the case where B is subsequently realigned; the result is that A remains aligned with C.) We say that A is *immediately aligned* with B but *ulti-*

mately aligned with C. If an object is not explicitly aligned with another object, we say that it is ultimately aligned with itself.

Every object is created as if according to some complete set of specification directives; if the program does not include complete specifications for the mapping of some object, the compiler provides defaults. By default an object is not aligned with any other object; it is ultimately aligned with itself. The default distribution is language-processor-dependent, but must be expressible as explicit directives for that implementation. Identically declared objects need not be provided with identical default distribution specifications; the compiler may, for example, take into account the contexts in which objects are used in executable code. The programmer may force identically declared objects to have identical distributions by specifying such distributions explicitly. (On the other hand, identically declared processor arrangements *are* guaranteed to represent "the same processors arranged the same way.")

Once an object has been created, it can be remapped by realigning it or redistributing an object to which it is ultimately aligned; but communication is required in moving the data around. Redistributing an object causes all objects then ultimately aligned with it also to be redistributed so as to maintain the alignment relationships.

Alignment is considered an *attribute* (in the Fortran 90 sense) of an array or scalar. Distribution is technically an attribute of the index space of the array. Sometimes we speak loosely of the distribution of an array, but this really means the distribution of the index space of the array, or of another array to which it is aligned. The relationship of an array to a processor arrangement is properly called the *mapping* of the array. (Even more technically, these remarks also apply to a scalar, which may be regarded as having an index space whose sole position is indicated by an empty list of subscripts.)

Sometimes it is desirable to consider a large index space with which several smaller arrays are to be aligned, but not to declare any array that spans the entire index space. HPF provides the notion of a TEMPLATE, which is like an array whose elements have no content and therefore occupy no storage; it is merely an abstract index space that can be distributed and with which arrays may be aligned.

By analogy with the Fortran 90 ALLOCATABLE attribute, HPF includes the attribute DYNAMIC. It is not permitted to REALIGN an array that has not been declared DYNAMIC. Similarly, it is not permitted to REDISTRIBUTE an array or template that has not been declared DYNAMIC.

Reference

1. Version 1.0 Draft, High Performance Fortran Language Specification, High Performance Fortran Forum, January 25, 1993, © 1992 Rice University, Houston, TX, copied with permission (CITI/CRPC, c/o Theresa Chatman, Rice University).

8.3   PARAGRAPH

Graphical visualization aids human comprehension of complex phenomena and large volumes of data. The behavior of parallel programs on advanced architectures is often extremely complex [1–5], and monitoring the performance of such programs can generate vast quantities of data. Therefore, it seems natural to use visualization [6, 7] to gain insight into the behavior of parallel programs so users can better understand them and improve their performance.

ParaGraph, a software tool that provides a detailed, dynamic, graphical animation of the behavior of message-passing parallel programs and graphical summaries of their performance was developed to meet this need.

For lack of a better term, "simulation" is used to mean graphical animation. This does not suggest that there is anything artificial about the programs or their behavior as ParaGraph portrays them. In effect, ParaGraph provides a visual replay of the events that actually occurred when a parallel program was run on a parallel machine.

ParaGraph has been used only for post-processing trace files created during execution and saved for later study. But its design does not rule out the possibility that data could arrive at the graphical workstation as the program executes.

However, there are major impediments to genuine real-time performance visualization. With the current generation of distributed-memory parallel architectures, it is difficult to extract performance data from the processors and send it to the outside world during execution without significantly perturbing the program being monitored. Also, the network bandwidth between the processors and the workstation and the drawing speed of the workstation are usually inadequate to handle the very high data-transmission rates that a real-time display requires. Finally, humans would be hard pressed to digest a detailed graphical depiction unfolding in real time. One of ParaGraph's strengths is that it lets you replay the same execution trace data repeatedly.

Some performance-visualization packages treat the trace file of events saved after a program executes as a static, immutable object to be studied by various analytical or statistical means. Such packages provide graphical tools designed for visual browsing of the performance data from various perspectives using scroll bars and the like.

ParaGraph adopts a more dynamic approach whose conceptual basis is algorithm animation. The trace file is seen as a script to be played out, visually reenacting the original action to provide insight into a program's dynamic behavior.

Both the static and dynamic approaches have advantages and disadvantages. Algorithm animation is good at capturing a sense of motion and change, but it is difficult to control the simulation's apparent speed. The static browser approach gives the user fine control over the speed at which the data are viewed (time can even move backward), but it does not provide an intuitive feeling for dynamic behavior.

The whole point of visualization is to aid understanding, so it is imperative that the visual displays be as intuitively meaningful as possible. The charts and diagrams should be aesthetically appealing and the information they convey should be self-evident. A diagram is not likely to be useful if it requires an extensive explanation, so the information it conveys should either be immediately obvious or easily remembered once learned.

The display's colors should reinforce the meaning of graphic objects and be consistent across views. Above all, the system must provide many visual perspectives, because no single view is likely to provide full insight into the behavior and data associated with parallel-program execution. ParaGraph provides more than 20 displays or views based on the same underlying trace data.

Software tools should relieve tedium, not promote it. ParaGraph has an interactive, mouse- and menu-oriented user interface so its features are easily invoked and customized.

Another important factor in ease of use is that the object under study (the parallel program) need not be modified extensively to obtain the visualization data. ParaGraph's input are trace files produced by the Portable Instrumented Communication Library (PICL) [8], which lets users produce trace data automatically.

Portability is important in two senses. First, the graphics package should be portable. ParaGraph is based on X Windows, and thus runs on many vendors' workstations. Although it is most effective in color, it also works on monochrome and grayscale monitors—it automatically detects which monitor type is in use.

Second, the package must be able to display execution behavior from different parallel architectures and parallel-programming paradigms. ParaGraph inherits a high degree of such portability from PICL, which runs on parallel architectures from many different vendors (including Cogent, Intel, N-Cube, and Symult).

On the other hand, many of ParaGraph's displays are based on the message-passing paradigm, so it does not support programs explicitly based on shared-memory constructs.

ParaGraph is distinguished from other visualization systems [9] in the following characteristics.

- The number of displays it provides. Although other packages provide multiple views, none we know of provides the variety of perspectives ParaGraph does. Some of ParaGraph's displays are original; others have been inspired by similar displays in other packages.

- Its portability among architectures and displays. ParaGraph is applicable to any parallel architecture having message passing as its programming paradigm, and ParaGraph itself is based on X Windows.

- The intuitive appeal and aesthetic quality of its displays.

- Its ease of use, attributable both to its interactive, graphic interface and to its use of PICL to provide the trace data without requiring the user to instrument the program under study.

- Its extensibility. ParaGraph lets users add new displays of their own design to the views already provided.

PICL runs on several message-passing parallel architectures [1]. As its name implies, it provides both portability and instrumentation for programs that use its communication facilities to pass messages among processors.

On request, PICL provides a trace file that records important execution events (such as sending and receiving messages). The trace file contains one event record per line, and each event record comprises an integer set that specifies the event type, time stamp, processor number, message length, and other similar information.

ParaGraph has a producer-consumer relationship with PICL: ParaGraph consumes the trace data PICL produces. Using PICL instead of a machine's native parallel-programming interface gives the user portability, instrumentation, and the ability to use ParaGraph to analyze behavior and performance.

These benefits are essentially "free" in that once you have implemented a parallel program using PICL, you do not have to change the source code to move it to a new machine (provided PICL is available on the new machine), and little or no effort is required to instrument the program for performance analysis.

On the other hand, because ParaGraph's dependence on PICL is solely for input data, ParaGraph could work equally well with any data source that has the same format and semantics. So other message-passing systems can be instrumented to produce trace data in ParaGraph's format, and ParaGraph's input routine can be adapted to different input formats. Indeed, ParaGraph has been used with communication systems other than PICL.

For meaningful simulation, the event's time stamps should be as accurate and consistent across processors as possible. This is not necessarily easy when each processor has its own clock with its own starting time and runs at its own rate. Also, the clock's resolution may be inadequate to resolve events precisely.

Poor clock resolution and/or synchronization can lead to what are called *tachyons* in the trace file—messages that appear to be received before they are sent (a tachyon is a hypothetical particle that travels faster than light). Tachyons confuse ParaGraph, because much of its logic depends on pairing sends and receives.

Because this possibility will invalidate some ParaGraph displays, PICL goes to considerable lengths to synchronize processor clocks and adjust for potential clock drift, so time stamps are as consistent and meaningful as possible. On some machines, PICL actually provides higher clock resolution than the system was supplied with.

Collecting trace data can add to overhead. PICL tries to minimize tracing perturbation by saving trace data in each processor's local memory, downloading them to disk only after the program has finished execution. Nevertheless, monitoring inevitably introduces extra overhead: In PICL, the clock calls necessary to determine the time stamps for the event records, plus other minor overhead, add a fixed amount (independent of message size) to the cost of sending each message.

Thus, the overhead added is a function of the frequency and volume of communication traffic; it also varies from machine to machine. In general, we believe that this perturbation is small enough that the program's behavior is not altered fundamentally. In our experience, the lessons we learn from the visual study of instrumented runs always improve the performance of uninstrumented runs.

ParaGraph supports command-line options that specify a host name for remote display across a network, forced monochrome display mode (useful if black-and-white hard copies are to be made from a color screen), or a trace-file name. The user can also specify (or change) the trace-file name during execution by typing the file name in an options menu. ParaGraph preprocesses the input trace file to determine some parameters (like time scale and number of processors) automatically, before the simulation begins; the user can override most of these values.

The initial ParaGraph display is a main menu of buttons with which the user controls the execution and selects submenus. Submenus include those for three display types, or families (utilization, communication, and tasks), for miscellaneous displays, and for options and parameters.

The user can select as many displays as will fit on the screen, and can resize displays within reason. It is difficult to pay close attention to many displays at once, but it is useful to have several simultaneous displays for comparison and selective scrutiny.

After selecting the displays, press start to begin the graphical simulation of the parallel program based on the trace file. The animation proceeds to the end of the trace file, but it can be interrupted with a pause/resume button. For even more detailed study, the step button provides a single-step mode that processes the trace file one event at a time.

You can also single out a time interval by specifying a starting and stopping time (the defaults are the beginning and ending of the trace file), or you can have the simulation stop at each occurrence of some event. You can restart the entire animation at any time by simply pressing the start button.

Some ParaGraph displays change in place as events occur, representing execution time from the original run with simulation time in the replay. Other displays represent execution time in the original run with one space dimension on the screen, scrolling (by a user-controllable amount) as simulation time progresses, in effect providing a moving window for viewing a static picture. No matter which time representation is used, all displays are updated simultaneously and synchronized with each other.

The relationship between simulation speed and execution speed is necessarily imprecise. The speed of the graphical simulation is determined primarily by the drawing speed of the workstation, which in turn is a function of the number and complexity of displays that have been selected. There is no way to make the simulation speed uniformly proportional to the original execution speed.

For the most part, ParaGraph simply processes the event records and draws the resulting displays as fast as it can. If there are gaps between consecutive time stamps, ParaGraph fills them in with a spin loop so there is at least a rough (if not uniform) correspondence between simulation time and execution time. Fortunately, close correspondence does not seem to be critical in visual performance analysis. What is most important is that the graphical replay preserve the correct relative order of events. Moreover, the figures of merit that ParaGraph produces are based on actual time stamps, not simulation speed.

Because ParaGraph's speed is determined primarily by the workstation's drawing speed, the number of displays you select can speed it up or slow it down. ParaGraph's speed is also affected by the displays' complexity and the type and amount of scrolling used.

ParaGraph is an interactive, event-driven program. Its basic structure is that of an event loop and a large switch that selects actions based on each event's nature. Menu selections determine ParaGraph's execution behavior, both statically (initial display selection, options, and parameter values) and dynamically (pause/resume, single-step mode).

ParaGraph has two event queues: a queue of X events produced by the user (mouse clicks, key presses, window exposures) and a queue of trace events produced by the program under study. ParaGraph must alternate between these queues to provide both a dynamic depiction of the program and responsive user interaction.

The X-event queue must be checked frequently enough to provide good responsiveness, but not so frequently as to degrade drawing speed. The trace-event queue must be processed as rapidly as possible while the simulation is active, but it need not be checked at all if the next possible event must be an X event (as happens before a simulation starts, after it finishes, when it is in single-step mode, or when it has been paused and can be resumed only by the user).

So ParaGraph's alternation between queues is not strict. Because not all event records PICL produces are of interest to ParaGraph, it fast-forwards through such uninteresting records before it rechecks the X-event queue. Also, ParaGraph checks the X-event queue with both blocking and nonblocking calls, depending on the circumstances, so workstation resources are not consumed unnecessarily when the simulation is inactive.

Most of the displays fall into one of three categories—utilization, communication, and task information—although some contain more than one type of information and a few do not fit these categories at all.

Utilization displays.   Utilization displays are processor-utilization displays that show how effectively processors are used and how evenly the computational work is distributed among them.

Communication displays.   Interprocessor-communication displays are helpful in determining communication frequency, volume, and overall pattern, and whether there is congestion in the message queues.

Task displays.   The displays thus far have depicted a number of important aspects that help detect performance bottlenecks. However, they contain no information about where in the parallel program the behavior occurs.

Task displays use information provided by the user and PICL to depict the portion of the program that is executing. The user defines tasks within a program by using special PICL routines to mark the beginning and end of each task and assign it a task number.

Source code for ParaGraph, as well as sample trace files for demonstrating its use, are available for free over Internet's Netlib software-distribution service [10].

REFERENCES

1. Heath, M.T., Ng, E., and Peyton, B.W., Parallel Algorithms for Sparse Linear Systems, *SIAM Review,* pp. 420–460, Sept. 1991.

2. Gantt, H.L., Organizing for Work, *Industrial Management,* pp. 89–93, Aug. 1991.

3. Kolence, K. and Kiviat, P., Software Unit Profiles and Kiviat Figures, *Performance Evaluation Rev,* pp. 2–12, Sept. 1973.

4. Morris, M.F., Kiviat Graphs: Conventions and Figures of Merit, *Performance Evaluation Rev,* pp. 2–8, Oct. 1974.

5. Lamport, L., Time, Clocks, and the Ordering of Events in a Distributed System, *Comm. ACM,* pp. 558–565, July 1978.

6. Tufte, E.R., *The Visual Display of Quantitative Information,* Graphics Press, Cheshire, CT, 1983.

7. Heath, Michael T., Visualizing the Performance of Parallel Programs, University of Illinois, National Center for Supercomputer Applications, 4157 Beckman Institute, Urbana, IL; and Etheridge, Jennifer A., Oak Ridge National Laboratory; IEEE Software, September 1991, p. 29 (IEEE, Washington, D.C.).

8. Geist, G.A., et al., PICL: A Portable Instrumented Communication Library, C Reference Manual, Tech. Report ORNL/TM-11130, Oak Ridge National Lab., Oak Ridge, TN, 1990.

9. Heath, M.T. and Etheridge, J.A., Visualizing Performance of Parallel Programs, Tech. Report ORNL/TM-11813, Oak Ridge National Lab., Oak Ridge, TN, 1991.

10. Dongarra, J.J. and Grosse, E., Distribution of Mathematical Software via Electronic Mail, *Comm. ACM*, pp. 403–407, May 1987.

Chapter 9

Applications

Applications are why we have the other elements, the hardware, the languages, the coding standards, the development environments, the diagnostic tools. The payoff from the applications is what provides the investment resources on which the other elements depend.

This chapter samples applications from environmental science, astronomy, earth resources, matrix operations for any discipline, tracking, and computational fluid dynamics. A diverse set by most measures, but not diverse in terms of computational intensity.

The applications have been chosen in part for their diversity. In the past there was a school of thought that held the view that parallel processing was suitable for only a narrow range of applications, that many problems and the algorithms used to solve them are not parallelizable. As languages, algorithms, and programming tools have improved, fewer and fewer hold this view.

9.1 TERRESTRIAL ECOSYSTEM MODEL ON THE CONNECTION MACHINE

The Terrestrial Ecosystem Model (TEM)[1] is a highly aggregated, process-based simulation model of carbon and nitrogen cycling in terrestrial ecosystems. It is grid-cell based, each grid cell being 0.5° latitude by 0.5° longitude, with no connections among adjacent grid cells. The TEM is designed to investigate interactions among terrestrial ecosystems and environmental variables at continental or global scales, with a maximum time step of 1 month. Environmental variables needed to run the model are: vegetation type, soil texture, potential and actual evapotransporation rates, solar irradiance, cloudiness, precipitation, temperature, and atmospheric CO_2 concentrations. These variable control C and N fluxes into and out of soils and vegetation, thereby influencing C and N masses in these compartments. The TEM was used in the cited paper to investigate the spatial and temporal distribution of net primary productivity in South America, under the assumption that all vegetation is mature and undisturbed by significant human land-use activities. The TEM and its implementation on the Paragon and the CM2 is described here.

The TEM contains five state variables: carbon in living vegetation (C_v), nitrogen in living vegetation (N_v), organic carbon in detritus and soils (C_s), organic nitrogen in detritus and soils (N_s), and available, inorganic soil nitrogen (N_{av}). The model was applied in that paper to the investigation of mature, natural ecosystems; parameters were defined such that state variables did not change from year to year for sites used to calibrate the model. For predictions at all other sites the model was run continuously until equilibrium conditions existed and all state variables remained virtually constant from year to year (i.e., for 100 years). The state variables do change from month to month according to differential inputs and losses driven by seasonal changes in climate.

$$\frac{dC_V}{dt} = GPP_t - R_{At} - L_{Ct} \tag{1}$$

$$\frac{dN_V}{dt} = UPTAKE_t - L_{Nt} \tag{2}$$

$$\frac{dC_S}{dt} = L_{Ct} - R_{Ht} \tag{3}$$

$$\frac{dN_S}{dt} = L_{Nt} - NETNMIN_t \tag{4}$$

$$\frac{dN_{AV}}{dt} = NINPUT_t - NETNMIN_t - NLOST_t - UPTAKE_t \tag{5}$$

where

$$t = \text{time (generally 1 month)}$$
$$C_V = \text{C in vegetation}$$
$$R_A = \text{Autotrophic respiration}$$
$$L_C = \text{C in litterfall, above and below ground}$$
$$N_V = \text{N in vegetation}$$
$$UPTAKE = \text{N uptake by vegetation}$$
$$L_N = \text{N in litterfall, above and below ground}$$
$$R_H = \text{Heterotrophic respiration (decomposition)}$$
$$C_S = \text{C in soil and detritus}$$
$$N_S = \text{Organic N in soil and detritus}$$
$$N_{AV} = \text{Available inorganic N in soil and detritus}$$
$$NETNMIN = \text{Net rate of mineralization of } N_S$$
$$NINPUT = \text{N inputs from outside ecosystem}$$
$$NLOST = \text{N losses from ecosystem}$$

Gross primary productivity (GPP) is defined in TEM as the total assimilation of CO_2-C by plants, excluding photorespiration. There are few existing measurements of the GPP of whole plant communities, so estimates are based on available information on net primary productivity and whole-plant respiration rates. Gross primary productivity is modeled as a function of the irradiance of photosynthetically active radiation, atmospheric CO_2 concentrations, moisture availability, mean air temperature, the relative photosynthetic capacity of the vegetation, and indirectly nitrogen availability.

For each time step:

$$GPP = (Cmax) \frac{PAR}{ki + PAR} \quad \frac{Ci}{kc + Ci} (TEMP)(Ac)(KLEAF) \qquad (6)$$

where

$\quad$ Cmax = the maximum rate of C assimilation by living vegetation under optimal environmental conditions $(gC \cdot m^{-2} \cdot mo^{-1})$

$\quad$ PAR = the irradiance of photosynthetically active radiation at canopy level $(cal \cdot cm^{-2} \cdot d^{-1})$

$\quad$ ki = the irradiance at which C assimilation proceeds at one-half its maximum rate

$\quad$ Ci = the concentration of CO_2 inside leaves $(\mu L \cdot L^{-1})$

$\quad$ kc = the internal CO_2 concentration at which C assimilation proceeds at one-half its maximum rate

TEMP, Ac, and KLEAF = unitless multipliers expressing the influences of air temperature, relative nutrient availability, and plant phenology, respectively, on GPP

The value of Cmax was defined for each vegetation type by adjusting its value during calibration of the model until predicted annual NPP matched the literature-based NPP estimate for the calibration site.

Increasing irradiance of photosynthetically active radiation (PAR) increases GPP hyperbolically, as has been demonstrated in a number of leaf studies. We estimated a mean value of 75 cal $\cdot$ cm^{-1} $\cdot$ d^{-1} for ki from published leaf studies, and assume it to be independent of vegetation type.

The influence of increasing atmospheric concentrations of CO_2 on GPP is assumed to follow Michaelis-Menton kinetics. It is assumed that CO_2 concentrations inside leaves (Ci) are directly proportional to atmospheric CO_2 concentrations (Ca) when stomata are fully open. Although Ci < Ca they are assumed to be equal when moisture is not limiting. This does not affect the equilibrium results but further refinement of this relationship will be required before investigating the effects of increasing atmospheric CO_2 concentrations.

Available information from CO_2-enrichment studies indicates that plant yields increase 24 to 50% with a doubling of atmospheric CO_2, given adequate nutrients and water. Assuming an intermediate value of 37%, the half-saturation constant kc was defined to be 200 $\mu L \cdot L^{-1}$. For model runs discussed here, atmospheric CO_2 was assumed to be a constant 340 $\mu L \cdot L^{-1}$.

Moisture limitations diminish CO_2 assimilation by modifying the conductance of leaves to CO_2 diffusion, and by directly modifying the biochemistry of photosynthesis. Only the first of these processes is considered in TEM. The best estimate of mean monthly moisture availability is the degree to which environmental demands for water are met by rainfall and available soil moisture, this relationship being expressed as the ratio of estimated actual evapotranspiration (EET) to potential evapotranspiration (PET). In TEM the relationship between CO_2 concentrations inside stomatal cavities (Ci) and in the atmosphere (Ca) is directly proportional to relative moisture availability.

$$G_v = 0.10 + (0.9 \text{ EET/PET}) \qquad (7a)$$

$$Ci = G_v \, Ca \qquad (7b)$$

where
G_v = a unitless multiplier that accounts for changes in conductivity owing to changes in moisture availability.

Decreasing moisture availability is assumed to increase stomatal closure, thereby decreasing the internal CO_2 concentrations. This relationship is based on the tight correlation found between transpiration rates and CO_2 assimilation.

Atmospheric CO_2 concentrations and moisture availability have interactive effects on potential GPP. The TEM predicts that doubling atmospheric CO_2 concentrations will result in larger increases in production when moisture stress is high than it will when moisture stress is low. This is consistent with empirical data.

Moisture stress also influences the phenology of vegetation, causing, for instance, leaf shedding. This latter factor is included in the phenology model, and further depresses GPP during dry seasons.

Temperature effects on GPP are assumed to be the same as measured temperature effects on net primary productivity. TEM models the temperature effect as a simple multiplier on potential GPP, with a maximum value of 1.0 (i.e., no effect) at the optimum temperature and lower values at all suboptimal temperatures. For each time step

$$\text{TEMP} = \frac{(T - Tmin)(T - Tmax)}{[(T - Tmin)(T - Tmax)] - (T - Topt)^2} \qquad (8a)$$

$$\text{TEMP} = 0.0 \text{ if TEMP} < 0.0 \qquad (8b)$$

where
TEMP = the temperature multiplier on GPP (no units)
  T = the mean monthly air temperature (°C)

Nitrogen availability influences GPP indirectly by influencing the relative allocation of effort toward C versus N uptake (Ac).

Phenological processes involved with leaf initiation and retention, enzymatic activity levels, and other processes that alter the ability of plants to utilize atmospheric CO_2 alter the rate at which photosynthesis proceeds, independent of the environmental relationships considered so far. A separate phenological model was developed that describes seasonal changes in the vegetation's capacity to assimilate C. This model simulates relative changes in the photosynthetic capacity of mature vegetation (KLEAF) from estimated actual evapotranspiration (EET) and the previous months photosynthetic capacity:

$$KLEAF = a\ (EET_t/EET_{max}) + b\ (KLEAF_{t-1}) + c \tag{9a}$$

$$KLEAF_t = 1.0 \qquad\qquad \text{if } KLEAF_t > 1.0 \tag{9b}$$

$$KLEAF_t = KLEAF_t/KLEAF_{max} \qquad \text{if } KLEAF_{max} < 1.0 \tag{9c}$$

$$KLEAF_t = min \qquad\qquad \text{if } KLEAF_t < min \tag{9d}$$

where
              t = the month
       EET_{max} = the maximum EET occurring during any month in that location
a, b, and c = regression-derived parameters
          min = a pre-established value below which the relative photosynthetic
                   capacity is not allowed to go
$KLEAF_{max}$ = the maximum predicted KLEAF from (9a) for a specific location.

Equations (9b) to (9d) normalize the predictions from (9a) to a maximum of 1.0 and a minimum value defined by the parameter min. Values of KLEAF as predicted from (9a–d) have no units, and are used as multipliers in the production equation (6).

Plant (autotrophic) respiration (R_A) is the total respiration (excluding photorespiration) of living vegetation, including all CO_2 production from the various processes of plant maintenance, nutrient uptake, and biomass construction. In TEM, R_A is the sum of maintenance respiration, Rm, and growth respiration, Rg

$$R_A = Rm + Rg \tag{10a}$$

TEM models maintenance respiration as a direct function of plant biomass (C_V), and assume that increasing temperatures increase maintenance respiration rates logarithmically with a Q_{10} of 2.0 over all temperatures:

$$Rm = Kr\,(C_V)\,e^{0.0693T} \qquad\qquad (10b)$$

where

Kr = the per-gram-biomass respiration rate of the vegetation at 0°C
      $(gC \cdot gC^{-1} \cdot mo^{-1})$

 T = the mean monthly air temperature (°C).

There is virtually no information available on whole-plant respiration rates in most ecosystems. Therefore the value of Kr was determined by calibrating TEM such that estimated total plant respiration (R_A) correctly matched the estimates of total autotrophic respiration for an oak-pine forest in New York, a *Liriodendron* forest in Tennessee and a tallgrass prairie in Oklahoma. We applied the mean Kr value derived from the two forested sites to all forests, the single grassland value to all grasslands, and a midway value to savannas, woodlands, and shrublands.

Growth respiration (Rg) is estimated to be 20% of the difference between GPP and Rm.

$$NPP_t' = GPP_t - Rm_t \qquad\qquad (10c)$$

$$Rg_t = 0.2\,NPP_t' \qquad \text{if } NPP_t' > 0.0 \qquad (10d)$$

$$Rg_t = 0.0 \qquad\quad \text{if } NPP_t' \le 0.0 \qquad (10e)$$

where

NPP' = the potential net primary production, assuming that the conversion effi-
        ciency of photosynthate to biomass is 100%. Respiration resulting from
        nutrient uptake is assumed to be part of Rm.

Net primary productivity (NPP) is defined as the difference between gross primary productivity and autotrophic respiration.

$$NPP = GPP - R_A \qquad\qquad (11)$$

The seasonality of NPP as defined by [11] may vary from observed seasonal changes in vegetation because TEM determines NPP when it occurs, not when growth from stored reserves occurs.

Carbon in litter production (L_C) is the total production of organic detritus by live plants, both above- and belowground, including all C losses in abscissed tissues, plant mortality, exudates, leachates, and herbivory. In the current version of TEM annual litter production is assumed equal to annual NPP. TEM models litter production (L_C) as a direct function of plant biomass (C_V).

$$KFALL = \frac{\text{(annual NPP)}}{12\ \text{(mean annual } C_V)} \qquad\qquad (12a)$$

$$L_{Ct} = C_{Vt}\,(KFALL) \qquad\qquad (12b)$$

Annual NPP and mean annual C_V are defined from literature sources. The parameter KFALL is assumed to be constant within each vegetation type.

Heterotrophic respiration (R_H) is the only loss of C from the detrital compartment (C_S). It is, in TEM, total C mineralization from detritus, and is therefore our estimate of total organic matter decomposition. Heterotrophic respiration is modeled as a function of soil C (C_S), mean air temperature (T), mean soil moisture, and the gram-specific decomposition constant Kd. For each month

$$R_H = Kd\ (C_S)\ e^{0.0693T}\ \text{MOIST} \tag{13}$$

As with plant maintenance respiration, TEM assumes that increasing temperatures increase the heterotrophic respiration rate with a Q_{10} of 2.0 over all temperatures. The value of Kd is determined by model calibration on a vegetation-specific basis.

MOIST is a function defining the influence of soil moisture on decomposition. Moisture is believed to influence decomposition via its influences on moisture availability at low soil moisture contents and on oxygen availability at high moisture contents. TEM defines these influences on a monthly basis from

$$B = \frac{M^{m1} - \text{Mopt}^{m1}}{\text{Mopt}^{m1} - 100^{m1}} \tag{14a}$$

$$\text{MOIST} = (0.8\ \text{Msat}^B) + 0.2 \tag{14b}$$

where
  M = mean monthly soil moisture (% saturation)
  m1 = a parameter defining the skewness of the curve
Mopt = the soil moisture content at which R_H is maximum (1.0)
Msat = a parameter that determines the value of MOIST when the soil pore
         space is saturated with water.

In (14b) the values of MOIST are normalized to range from 0.2 to 1.0 without units. Different curves are defined for each of the five soil texture classes considered.

The specific relationships between relative decomposition rate and soil moisture are based on the rule of thumb that maximum decomposition rates occur when soils are 50 to 80% saturated with water. In soils of different water-holding capacities this occurs when the soil volume is about 15% air; this point defines our optima. TEM also sets the rate of decomposition at a soil moisture tension of 1.5 MPa to be 30% of the maximum rate, and the decomposition rate at saturation to be 60 to 80% of its maximum rate. These values typify the approximate means of a variety of laboratory and field decomposition studies. Finally, TEM assumes that the minimum relative decomposition rate with respect to soil moisture is 0.2. This latter assumption is based in part on the effect of scaling soil moisture across large areas, but also indicates that decomposition may be little influenced by soil moisture under dry conditions.

Net Ecosystem Productivity (NEP) is defined as the net rate of C accumulation by the ecosystem

$$NEP = GPP - R_A - R_H \tag{15}$$

Annual NEP is assumed to be zero under the equilibrium conditions discussed here.

Nitrogen fluxes and carbon-nitrogen interactions are also modeled in TEM; see the reference.

GE's Advanced Technology Laboratories (ATL), with cooperation from the University of New Hampshire (UNH), ported the Terrestrial Ecosystem Model to the Connection Machine (CM-2). The ported code was 1400 lines in Fortran written for a Prime computer. The test data corresponds to South America; 0.5 deg $\times$ 0.5 deg cells cover South America with 6192 cells.

Initially the sequential code was implemented on a Sun 4/470 using both the standard Sun Fortran compiler and the CMFortran compiler from Thinking Machines Corp.

The parallelization strategy employed used serial integration of state variables within each grid cell and parallel computation of all grid cells. Other strategies, for example, parallel integration of state variables, were examined but not employed because of excessive delays associated with processor to processor communications.

The ATL Connection Machine is a 16K CM-2 with 256 K bits of memory per processor and 32-bit floating point accelerator chips. The host is a Sun 4/470; on-line mass storage is provided by a 5 Gigabyte DataVault. ATL has observed a peak performance of 2 GigaFLOPS.

TEM was also modified for execution on MIMD computers. The parallelization approach used was a master/slave arrangement between the host computer and nodes of the parallel processor. The computations associated with a grid cell are performed on the nodes. The host reads the input data for a grid, passes the data to the next available node, writes the results returned from the node (for its previous grid cell) to the output file, and continues with the next grid cell.

This approach was implemented for the Intel Touchstone Gamma (iPSC/860) computer. Development and debugging was performed on a Sun/4 workstation using an iPSC/860 simulator provided by Intel. The original TEM code was divided into two parts for the host and node programs. New code for coordination between the host and nodes consists of approximately 50 lines of FORTRAN, plus subroutines to copy data to and from a message buffer. The entire conversion effort took approximately 2 man-days.

The parallel TEM code was executed on a 64-node Touchstone Gamma provided by Intel, using Internet access. A 32-node subcube was used for benchmarking the code. For 6192 grid cells, a 600-month simulation executed in 930 seconds. It should be noted that, in this implementation, file I/O is performed on the host's file system. It is expected that the execution time will be much shorter when the data files are resident on the fast access Concurrent File System. This expectation is justified since additional experiments indicate that only about 12 processing nodes can be adequately serviced by the host I/O capability given the amount of data that must be input for each grid cell.

The projections in Table 9-1 assume that the I/O time for the CM-2 implementation is linear with the number of grid cells in the model and the processing time is constant until the number of cells exceeds the number of processing elements. That is, 65,536 cells on a 65,536 processor CM-2 still requires 507.3 sec processing time. One additional cell will double the required processing time. The Sun execution time is estimated to be totally linear since each grid cell is processed sequentially.

The size of a global model covering land area only is estimated below based on the ratio of land areas to South America, for which there are 6192 grid cells.

Location	Cells
Africa	12,120
Asia	18,000
Australia	3073
North America	9760
South America	6192
Europe	3960
Total	53,105

Additional projections are given in Table 9-2 that show the impact of higher temporal and spatial resolution in the model. An order of magnitude increase in both spatial and temporal resolution would result in the 1000x processing. In addition, UNH estimates that the inclusion of nitrogen feedback into the TEM requires approximately 5x processing. The results in Table 9-2 point out that higher

TABLE 9-1

TIME ESTIMATES FOR BENCHMARKED TEM VERSION APPLIED TO LARGER PROBLEMS ON LARGER CMS				
# Cells	CM Size	Sun 4/470 Time (sec)	CM-2 Time (sec)	Speedup
6192	8192	9449	538	17.6
16,384	16,384	25,000	589	42.4
65,536	65,536	100,000	832	120.2

TABLE 9-2

Estimated CM-2 Computation Time Projected for TEM Applied to Global Land Area			
Model	**16 K CM-2**	**64 K CM-2**	**teraFLOP SIMD Proc**
Baseline Model 50 km × 50 km cells, 30 day timestep, 53,105 cells	38 min	13 min	3 sec
Baseline Model with Nitrogen Feedback	3 hr	47 min	11 sec
High Resolution Model 5 km × 5 km cells, 3 day timestep, 5,310,500 cells	22 days	8 days	45 min
High Resolution Model with Nitrogen Feedback	98 days	27 days	2.5 hrs

resolution alone is sufficient to require teraFLOP computational capability. The integration of hydrology and atmospheric models will even further increase the computing requirements.

However, to approach the desired level of similarity to the real world, two more features need to be incorporated into TEM. The first is cell interaction, in which moisture and gases can migrate across cell boundaries. This will require topographic and weather information, although prevailing wind historical data can be used instead of detailed weather information during development.

The second additional feature is atmospheric–ocean interactions. This is particularly so for oceanic algae photosynthesis.

With these features added, TEM is estimated to have runtimes increased by two orders of magnitude. The High Resolution Model with Nitrogen Feedback is projected to have a runtime of 250 hours on a teraflop computer.

REFERENCE

1. Raich, J.W., et al., Potential Net Primary Productivity in South America: Application of a Global Model. *Ecological Applications* 1(4), 1991, pp. 399–429.

9.2   PHASE-DIVERSE SPECKLE IMAGING

Fine-resolution imaging of extended objects through atmospheric turbulence can be accomplished with stellar speckle interferometry, which requires the collection of many image frames of the same object. The exposure time for each frame must be short enough ($\approx$10 msec) that the evolving atmosphere can be regarded as frozen during the exposure. Labeyrie observed that these short-exposure images have a speckle texture with features comparable in size to the diffraction-limited resolution of the system [1]. Thus, diffraction-limited information regarding the object could be encoded in these short-exposure images or *specklegrams*. The data-collection scheme for stellar speckle interferometry is illustrated in Figure 9.2-1(A). Current stellar speckle interferometric approaches, including Labeyrie's method, perform an averaging of the data in the process of individually estimating the Fourier modulus and the Fourier phase of the object. The final estimated object is constructed by combining the Fourier modulus and phase estimates and performing an inverse Fourier transform. In addition, a calibration procedure must be performed using images of an unresolved object through atmospheric turbulence having the same statistics [2]. This calibration step is particularly problematic for imaging solar scenes, since isolated unresolved objects are not available in this case [3, 4].

Reconstruction of an object from a sequence of specklegrams could be substantially improved if estimates for the individual PSFs (point spread functions) for each specklegram were available. This is the rationale behind *deconvolution with wavefront sensing* [5, 6], which employs a wavefront sensor to estimate the PSFs for each specklegram.

The method of *phase diversity* [7–11], first proposed by Gonsalves, may be regarded as an indirect wavefront sensor since phase aberrations are estimated from image data. The technique requires the collection of two or more images. One of these images is the conventional focal-plane image that has been degraded by the unknown aberrations. Additional images of the same object are formed by perturbing these unknown aberrations in some known fashion, thus creating its phase diversity, and reimaging. This can be accomplished with very simple optical hardware. For example, a simple beam splitter and a second detector array, translated along the optical axis, further degrades the imagery with a known amount of defocus, as illustrated in Figure 9.2-1(B). This particular type of phase diversity (defocus) has been referred to by some researches as the focal-volume technique [12–14].

A novel data-collection and processing approach, called *phase-diverse speckle imaging* (PDSI) [15, 16], that makes use of both speckle-imaging and phase-diversity concepts is described here. The data collection consists of one conventional specklegram and at least one additional specklegram having phase diversity for each of multiple atmospheric realizations, as illustrated in Figure 9.2-1(C). The principle of maximum-likelihood can be used to jointly estimate the object and aberrations, given this data set. This approach is appealing for the following reasons.

1. The optical hardware needed is simple.

2. The method is robust to systematic errors because it relies on the external object as a reference.

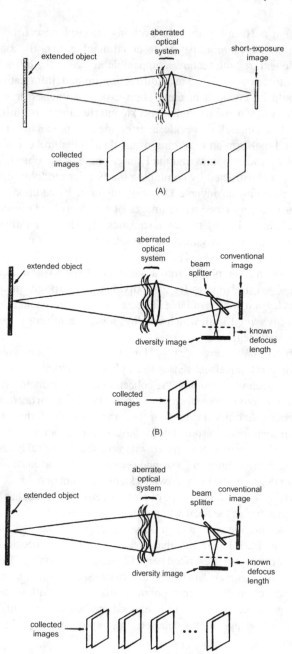

Figure 9.2-1.   Approaches to imaging through the atmosphere. (A) Stellar
speckle interferometry; (B) Phase diversity; (C) Phase-diverse
speckle imaging.

3. No speckle calibration procedure is required.

4. There is no averaging of the data in which information is lost.

5. The object estimation is direct, with no intermediate estimates (the Fourier modulus and phase are jointly estimated).

An estimation-theoretic approach to accomplish the *joint* estimation of the object (common to all collected frames) and the phase aberrations for each atmospheric realization is used. An incoherent imaging model is constructed that captures the dependence of the noiseless imagery on the object and the optical system, including atmospheric aberrations. A noise model is then introduced to account for detector or photon noise that is present in the detected imagery. It is then possible to write down a probability law for the collected imagery, given the object and the system aberrations. The method of maximum likelihood is used to construct an estimator for the object and aberrations.

The phase-diverse speckle imaging concept (Figure 9.2-1(C)) is demonstrated here with a simulation result [18–20]. A portion of a Solar Optical Universal Polarimeter (SOUP) image with 0.1 arcsecond resolution was down-sampled by a factor of 5 onto a 64 × 64 grid and used as the *ground truth object* for the simulation. This image provides realistic features of interest but no attempt was made to match the intrinsic resolution of this image to the resolution predicted by the simulation. A telescope with a 1 m aperture and a 20 cm obscuration was used as a system model for collecting two images of the solar scene for each atmospheric realization: a conventional (focal-plane) image and a diversity image with a known amount of defocus. The telescope aperture was simulated on a 32 × 32 grid. A series of 5 atmospheric realizations was used to generate a total of 10 simulated data frames. The atmospheric aberrations were simulated using a Kolmogorov spatial power spectrum [17] with a turbulence strength equivalent to $\underline{r}_0 = 12$ cm seeing.

The conventional and defocused short-exposure images for a representative atmospheric realization are shown in Figure 9.2-2(A) and (B), respectively. Photon noise was added to each of these images such that the photon count at the peak of each is about 100,000, a reasonable fraction of the full-well depth of a typical CCD camera. The ideal (infinite photons) diffraction-limited (unaberrated) image of this object is shown in Figure 9.2-2(C), and it is clear that the conventional image in Figure 9.2-2(A) has been significantly degraded by the atmospheric phase aberration. The diversity image is degraded further, albeit by a known amount of two waves of defocus. For ease of comparison, the same bias and scale were used for all the images in Figure 9.2-2.

The conventional image in Figure 9.2-2(A), the diversity image with known defocus in Figure 9.2-2(B), and the four additional image-pairs corresponding to the other atmospheric realizations constitute the data used to drive the maximum-likelihood estimation of the object and the atmospheric phase aberrations. Beginning with the naive estimates that there are no phase aberrations in any of the five

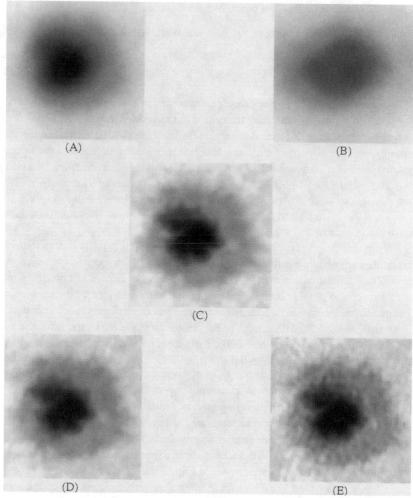

Figure 9.2-2.   Simulation demonstration of phase-diverse speckle imaging.
(A) Conventional short-exposure image as observed through
a single atmospheric realization (100,000 photons at peak);
(B) Defocused image using the same atmospheric realization
(97,000 photons at peak); (C) Ideal, diffraction-limited image
(infinite photons); (D) Estimated diffraction-limited image using
phase-diverse speckle imaging for five atmospheric realizations
(10 total images); (E) Estimated diffraction-limited image with
conventional phase-diversity using only (A) and (B).

collections and that the object is constant-valued, the phase-diversity algorithm
attempts to find the aberrations and the object that are consistent with the 10
measured images. The method of conjugate gradients was used to maximize the

log-likelihood function that is defined in a multidimensional parameter space. Both the object and the atmospheric aberrations are estimated on a point-by-point basis. That is, an estimate is made of every object pixel and every phase aberration sample over the aperture. The 64×64 pixel object and five 32×32 pupil arrays with obscuration (one for each atmospheric aberration) combine to roughly 7000 parameters that are simultaneously estimated.

The estimated diffraction-limited image shown in Figure 9.2-2(D) was obtained by convolving the unaberrated telescope point-spread function (PSF) with the phase-diverse speckle object estimate. This has the effect of suppressing high spatial-frequency artifacts in the object estimate. The estimate in Figure 9.2-2(D) is in excellent agreement with Figure 9.2-2(C), the diffraction-limited image that would be collected if atmospheric aberrations were eliminated altogether. We find it remarkable that the fine-resolution detail in Figure 9.2-2(D) can be derived from imagery such as that shown in Figure 9.2-2(A) and (B).

The phase-diverse speckle estimate (Figure 9.2-2(D)) can be compared with Figure 9.2-2(E), the estimate of the diffraction-limited image using conventional (single atmospheric realization) phase-diversity (Figure 9.2-1(A)). This image was obtained (again via convolution with the unaberrated system PSF) from the object estimated using only the images in Figure 9.2-2(A) and (B). The advantage of the phase-diverse speckle technique is clearly demonstrated here. Specifically, the reconstruction from five atmospheric realizations (Figure 9.2-2(D)) is both sharper and less noisy than the reconstruction from a single atmospheric realization (Figure 9.2-2(E)). Whereas collecting images from multiple atmospheric realizations increases the number of aberration functions that must be estimated, each collected image is derived from a common object, suggesting the accumulation of object information. As the number of short-exposure images increases, the ratio of the number of parameters to estimate to the number of data samples available decreases and the fidelity of the estimates is significantly improved.

Although the hardware needed to collect PDSI data is simple and inexpensive, the PDSI algorithm that performs the object estimation is computationally intensive. An overview of the processing is shown in Figure 9.2-3. The philosophy is to trade away collection hardware complexity for increased computation, thus reducing the need for expensive hardware solutions to the imaging-through-turbulence problem (such as additional space-based telescopes). This trade is particularly appealing with the increased availability of high performance computers, such as the Maui High Performance Computing Center (MHPCC) IBM SP1.

ERIM has implemented a parallelized version of the Phase-Diverse Speckle Imaging simulation (called apd) on the Maui High Performance Computing Center's SP1 Parallel Computer using parallel Linda, and for comparison, on a network of workstations also using SCA's Linda package for networks. The inputs to the computationally intensive process are multiple pairs of images of a common object; in the parallel version each image is assigned to its own node.

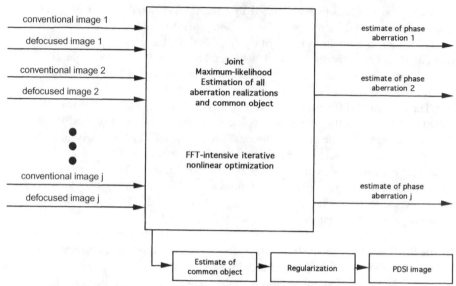

Figure 9.2-3.   Phase-Diverse Speckle Imaging Processing.

The apd has been parallelized using a strategy that attempts to minimize the number of bytes that have to be sent back and forth between the "master" and "slave" nodes. Nonetheless, the largest transfers for a 64×64 simulation are 65 Kbytes. The SP1 affords the advantage of using a high-speed interface for data transfers, whereas the ERIM network Linda version relies on a local ethernet to transfer data back and forth. The difference in transfer rates makes a substantial difference in overall run time, as shown in the following. Another disadvantage of the network Linda package is that it uses a heterogeneous network of computers with various speeds, and it will only do as well as the slowest of the machines used. In the timing information provided in the following, a combination of Sparc 10s and Sun 4s were available. The Sun 4s are about two-thirds as fast as the Sparc 10s.

Using Linda v 2.5.2, a version of apd that ports directly to the SP1 on Maui was compiled. Not only does it port directly, but Linda takes care of distributing processes to the various SP1 nodes in such a way that the program is invoked with a single command line with a single additional switch that requests a particular number of nodes.

For the case study presented in the following (Table 9-3), five pairs (10 images) with 64×64 pixels each were used as input and the optimization process was allowed 200 iterations.

There is a disadvantage in parallelization when ERIM's network Linda is used. The bottleneck is in the data transfer on the ethernet. The improved performance provided by the SP1 is primarily owing to the high-speed interface for

TABLE 9-3

PDSI*		
# Nodes	ERIM Network Linda	SP1
1	9273 s (single Sparc 10)	9631 s (single RS/6000)
2	N/A	5154 s (1.9x)
5	9681 s (0.96x)	2886 s (3.3x)
10	11,330 s (0.82x)	2493 s (3.9x)

*64 × 64 timing results (speed-up factor over single node in parentheses).

data transfers. This improvement will be even more pronounced with larger images and when operating on more image pairs (needed for fainter objects such as satellites and stellar objects).

With the SP1 parallelization, the ideal improvement would be linear, by a factor roughly equal to the number of nodes used. When two nodes are used, the expected speed-up of roughly a factor of two was achieved. As more nodes are added, however, a drop-off from the theoretical ideal occurs, indicating an overhead factor that is related to the number of nodes being used.

This implementation is preliminary; with little investment it is allowing simulation studies on this novel imaging modality with much greater efficiency. The parallelization and both network and SP1 evaluations required about 1 man-week of effort.

REFERENCES

1. Labeyrie, A., Attainment of diffraction limited resolution in large telescopes by Fourier analyzing speckle patterns in star images, *Astron. Astrophys.* Vol. 6, pp. 85–87, 1970.

2. Dainty, J.C., Stellar speckle imaging, in *Topics in Applied Physics: Laser Speckle and Related Phenomena*, Dainty, J.C., ed. New York, Springer-Verlag, 1984.

3. von der Lühe, O., High spatial resolution techniques, in *Solar Observations: Techniques and Interpretations*, Sánchez, F., Collados, M., and Vázquez, M., eds. Cambridge University Press, New York, 1992.

4. von der Lühe, O., Speckle imaging of solar small scale structure: 1. Methods, accepted for publication in *Astronomy and Astrophysics*.

5. Primot, J., Rousset, G., and Fontanella, J.C., Deconvolution from wave-front sensing: a new technique for compensating turbulence-degraded images, *J. Opt. Soc. Am.* Vol. A7, pp. 1598–1608, 1990.

6. Gonglewski, J.D., Voelz, D.G., Fender, J.S., Dayton, D.C., Spielbusch, B.K., and Pierson, R.E., First astronomical application of postdetection turbulence compensation: images of ∝ Aurigae, ν Ursae Majoris, and ∝ Geminorum using self-referenced speckle holography, *Appl. Opt.* Vol. 29, pp. 4527–4529, 1990.

7. Gonsalves, R.A. and Childlaw, R., Wavefront sensing by phase retrieval, in *Applications of Digital Image Processing III*, Tescher, A.G., ed., *Proc. Soc. Photo-Opt. Instrum. Eng.* Vol. 207, pp. 32–39, 1979.

8. Gonsalves, R.A., Phase retrieval and diversity in adaptive optics, *Opt. Eng.* Vol. 21, pp. 829–832, 1982.

9. Paxman, R.G. and Fienup, J.R., Optical misalignment sensing and image reconstruction using phase diversity, *J. Opt. Soc. Am.* Vol. A5, pp. 914–923, 1988.

10. Paxman, R.G. and Crippen, S.L., Aberration correction for phased-array telescopes using phase diversity, in Digital Image Synthesis and Inverse Optics, *Proc. Soc. Photo-Opt. Instrum. Eng.* Vol. 1351, pp. 787–797, 1990.

11. Paxman, R.G., Schulz, T.J., and Fienup, J.R., Joint estimation of object and aberrations by using phase diversity, *J. Opt. Soc. Am.* Vol. A9, pp. 1072–1085, 1992.

12. Högbom, J.A., On the intensity distribution over the focal volume, in *High Spatial Resolution Solar Observations*, Proceedings of the 10th Sacramento Peak Summer Workshop, Sunspot, New Mexico, 1988.

13. Högbom, J.A., Reconstruction from focal volume information, in *Solar and Stellar Granulation*, Rutten, R.J. and Severino, G., eds. Kluwer Academic Publishers, Hingham, MA, 1989.

14. Restaino, S.R., Wavefront sensing and image deconvolution of solar data, *Appl. Opt.* Vol. 35, pp. 7442–7449, 1992.

15. Paxman, R.G. and Seldin, J.H., Fine-resolution imaging of solar features using phase-diverse speckle imaging, in *Real Time and Post-Facto Solar Image Correction* 13th Sacramento Peak Summer Workshop, Sunspot, NM, September 1992.

16. Paxman, R.G., Schulz, T.J., and Fienup, J.R., Phase-diverse speckle interferometry, in *Topical Meeting on Signal Recovery and Synthesis IV*, Technical Digest Series 11 (Optical Society of America, Washington DC, 1992), New Orleans, LA, April 1992.

17. McGlamery, B.L., Computer simulation studies of compensation of turbulence degraded images, *SPIE/OSA Image Processing* Vol. 74, pp. 225–233, 1976.

18. Schulz, T.J., Image recovery from atmospherically degraded images with unknown point-spread functions, in *Topical Meeting on Signal Recovery and Synthesis IV*, Technical Digest Series 11 (Optical Society of America, Washington DC, 1992) New Orleans, LA, April 1992.

19. Schulz, T.J., Multi-frame blind deconvolution of astronomical images, accepted for publication in *J. Opt. Soc. Am.*

20. Paxman, Richard G. and Seldin, John H., *Fine-Resolution Astronomical Imaging Using Phase-Diverse Speckle*, Proceedings of Workshop on Wavefront Supported by Post-Facto Image Correction, Anderson, Torben, ed., Riso, Roskilde, Denmark, November 1992.

9.3   MULTISPECTRAL IMAGE PRODUCT GENERATION

In 1991, General Electric's Advanced Technology Laboratories in Moorestown, NJ conducted a cost-effectiveness study comparing the Connection Machine CM-2 (from Thinking Machines Corp.) with a Sun 4/470 in the generation of multispectral image products. The study was intended to examine the benefits of massively parallel processing for image transformations. The study was motivated by the realization that future space-borne imaging sensors will produce enormous volumes of data needing transformation into product formats on a production basis. For example, HIRIS, the high resolution imaging spectrometer planned for NASA's Earth Observation Satellite, may produce in 192 spectral channels, $1.2 \times 10^{**}9$ pixels per day per channel. Another sensor, MODIS-N, with 36 spectral channels is expected to generate $1.1 \times 10^{**}10$ pixels per day per channel.

Both geometric and radiometric transformations were applied. The geometric transformation chosen was the affine projection. This transformation produces a version of an image as if obtained from a different point of view. For example, images obtained from an oblique point of view can be made to appear as if obtained at nadir, that is, from directly overhead. This plan view vantage point version is particularly useful for registering imagery to other images or maps, or for correlation of multisensor imagery.

The affine transformation equations employed were

$$x' = (a1 * x + a2 * y + a3)/(a7 * x + a8 * y + 1)$$

$$y' = (a4 * x + a5 * y + a6)/(a7 * x + a8 * y + 1)$$

The a_i, the eight coefficients in the affine transformation equations, are determined through a least squares fitting (LSF) process prior to the application of the geometric transformation described in the following. The LSF depends on selecting a minimum of 5 control point pairs, that is, $N = 5$ or more sets of $(x, y; x', y')$ values. In registering an image to a map, for example, the x, y coordinates of the control points in the map would be matched with the corresponding x', y' coordinates in the image. The primed coordinate values are arranged into a vector as shown in the following.

$$
P = \begin{matrix}
x_1' \\
x_2' \\
\cdot \\
\cdot \\
\cdot \\
x_N' \\
y_1' \\
y_2' \\
\cdot \\
\cdot \\
\cdot \\
y_N'
\end{matrix}
$$

The a_i coefficients are formed into a vector as well.

$$
A = \begin{matrix}
a_1 \\
a_2 \\
a_3 \\
\cdot \\
\cdot \\
\cdot \\
a_8
\end{matrix}
$$

Then the primed and unprimed coordinates are used to construct a matrix, U, so that the matrix equation, $P = UA$, is equivalent to the affine transformation equations.

$$
U = \begin{matrix}
x_1 & y_1 & 1 & 0 & 0 & 0 & -x_1x_1' & -y_1x_1' \\
x_2 & y_2 & 1 & 0 & 0 & 0 & -x_2x_2' & -y_2x_2' \\
\cdot & \cdot & \cdot & \cdot & \cdot & \cdot & \cdot & \cdot \\
x_N & y_N & 1 & 0 & 0 & 0 & -x_Nx_N' & -y_1y_1' \\
0 & 0 & 0 & x_1 & y_1 & 1 & -x_1y_1' & -y_1y_1' \\
0 & 0 & 0 & x_2 & y_2 & 1 & -x_2y_2' & -y_2y_2' \\
\cdot & \cdot & \cdot & \cdot & \cdot & \cdot & \cdot & \cdot \\
0 & 0 & 0 & x_N & y_N & 1 & -x_Ny_N' & -y_Ny_N'
\end{matrix}
$$

To solve the matrix equation, $P = UA$, for A, both sides are premultiplied by U transposed, that is, U^T, so $U^TP = U^TUA$. This is then premultiplied on both sides by the inverse of (U^TU), that is, $(U^TU)^{-1}$, so $(U^TU)^{-1}U^TP = (U^TU)^{-1}(U^TU)A$. Since $(U^TU)^{-1}(U^TU)$ is the identity matrix, $A = (U^TU)^{-1}U^TP$.

where
(x,y) = pixel coordinates in the new (transformed) image
$(x'y')$= pixel coordinates in the old image.

In applying this affine transformation there are two ways to set the value of the pixel at (x,y): nearest neighbor, which takes the value of the old image pixel closest to (x',y'), and interpolation, which computes a value based on combining values from multiple old image pixels near (x',y'). Bilinear interpolation uses a 2×2 set of old image pixels, biquadratic interpolation uses a 3×3 set of old image pixels and bicubic interpolation uses a 4×4 set. In this study both nearest neighbor and bicubic interpolation were examined.

The radiometric transformation employed was a convolution using a 7×7 kernel. This transformation is rather general and is often employed to remove the effects of a point spread function, to enhance edges, or to reduce noise. In this case the kernel was not assumed to be symmetric, and not assumed to be decomposable into one-dimensional convolutions.

The data employed by this study consisted of Coastal Zone Color Scanner (CZCS) data from NASA; a frame consisted of 1968 pixels by 970 pixels, 8 bits per pixel. The images were transformed in smaller sections, measuring 512×512 pixels.

The nearest neighbor version of the affine transformation used fixed point computation; floating point (single precision, 32-bit) computation was used for calculation of the affine transformation with bicubic interpolation and the convolution.

The implementations compared were a C version on a Sun 4/470 with 32 MB of RAM and a C* version on a Connection Machine CM-2 with 16,384 processors, 512 MB memory, and 32-bit floating point accelerator chips. The results are shown in Table 9-4. The Nearest Neighbor column applies to the case of the affine transformation using the nearest neighbor rule, without convolution. The bicubic interpolation column applies to the case of the affine transformation using the bicubic interpolation rule, without convolution. The Plus 7×7 Convolution column applies to the case of the affine transformation using the bicubic interpolation rule together with the 7×7 convolution.

The cost performance ratio shown on the bottom row indicates that the nearest neighbor case is not cost effective on the CM-2 because for every dollar spent on this application using the CM-2, one could obtain those results on the Sun 4/470 for .50¢. However, as the computation becomes more complex, the massively parallel approach improves in cost effectiveness. The affine transformation with bicubic interpolation has a cost-effectiveness ratio of 1.481, indicating that a dollar's worth of processing on the CM-2 would cost $1.48 on the Sun 4/470; adding the convolution drives the cost-effectiveness ratio even higher, to 2.54, making the massively parallel approach more attractive.

Table 9-5 shows the estimated processing time, using the times from Table 9-4, per day per channel for anticipated HIRIS and MODIS-N data sets, for the affine transformation with bicubic interpolation and a 7×7 convolution.

TABLE 9-4

RESULTS			
	Nearest Neighbor	**Bicubic Interpolation**	**Plus 7 × 7 Convolution**
SUN 4/470	2.833 sec	19.949 sec	44.332 sec
16 K CM	0.200 sec	0.473 sec	0.613 sec
Speed Up	14.16 ×	42.18 ×	72.32 ×
Cost Performance Ratio*	0.498	1.481	2.540

*(SUN time) × (SUN cost)
 (CM time) × (CM cost)

TABLE 9-5

IMAGE TRANSFORMATION STUDY		
Extrapolated Processing Times Based on affine transformation with bicubic interpolation and 7 × 7 convolution Estimated processing time per day per channel		
	HIRIS* (hrs)	**MODIS-N** (hrs)**
SUN 4/470	5.91	541.8
16 K CM	0.82	7.49

* HIRIS has 192 channels.
** MODIS-N has 36 channels.

9.4   ALGORITHM-BASED FAULT TOLERANCE ON THE CONNECTION MACHINE

With the increase in computer hardware complexity comes an increase in the chances of hardware failure. Such a failure can destroy the results of lengthy calculations, such as those performed in signal processing, image processing, and numerical linear algebra. Thus some level of fault tolerance is required to ensure that the computed results are not polluted by a hardware fault, or at least to report that the integrity of the results is questionable.

The traditional approach to fault tolerance has been through the use of hardware-based or software-based fault tolerance, which have a 100 to 300% overhead in hardware or time redundancy [1]. Recently, the approach of algorithm-based fault tolerance (ABFT) has been developed [1] to provide fault tolerance with a much smaller overhead than traditional approaches.

ABFT techniques have been applied to signal and image processing algorithms (e.g., FFT [2–5], FIR [2, 5], and convolution [2, 5, 6]), matrix operations (e.g., matrix multiplication [1, 2, 7], LU factorization [1, 2, 5, 8–10], QR factorization [2, 5, 8–11], and conjugate gradient [12]), and solution of Laplace's equation [13].

In this section, the use of ABFT on the Connection Machine is examined [14]. The approaches developed primarily by Huang and Abraham [1] and Luk and Park [8] to detect, locate, and correct an error caused by a single fault are reviewed. Extensions of these approaches to handle multiple faults and determine how many faults can be handled by each method are then considered [15]. An ABFT approach that is appropriate for the Connection Machine and how it can be implemented on the Connection Machine are developed. Finally, the results of numerical computations on the Connection Machine to demonstrate the overhead cost actually achieved are presented.

Given an $N \times M$ matrix A called the *information matrix*, the matrix will be *encoded* into a *checksum matrix* using an *encoding* (or *weight*) *vector*, w. Here, the information matrix A is any matrix one intends to use by addition, multiplication, inversion, transposition, or any other operation in the course of executing an algorithm. To obtain algorithm based fault tolerance the checksum matrix is used rather than the information matrix, because the checksum matrix contains all the information carried by the information matrix plus the checksum entries with which the information matrix has been augmented. These checksum entries allow the detection and correction of any matrix entry that may have become incorrect due to a hardware fault. A commonly used encoding vector is $e = (1,1,\cdots,1)^T$. A *checksum column* is a vector Aw; a *checksum row* is a vector $w^T A$. When $w \neq e$, the checksum column (row) is said to be a *weighted checksum column (row)*. The *full checksum matrix* is the information matrix appended with the unweighted checksum column and unweighted checksum row, that is,

$$\begin{bmatrix} A & Ae \\ e^T A & e^T Ae \end{bmatrix}$$

The *row weighted checksum matrix* is the information matrix appended with the unweighted checksum column and a weighted checksum column, that is,

$$[A \; Ae \; Aw]$$

There are three distinct processes in ensuring the results of a matrix operation are correct. *Detection* is the process of determining whether or not an erroneous value exists in the result. *Location* is the process of finding which value(s) are erroneous. *Correction* is the process of replacing the erroneous value(s) with the correct result.

Two approaches to detecting, locating, and correcting a single error in a checksum matrix are prevalent in the literature, one using the full checksum matrix and the other using the row weighted checksum matrix.

Huang and Abraham [1] suggested the use of a full checksum matrix to detect, locate, and correct a single error. If a full checksum matrix has a single erroneous element, that error can be detected through an inconsistent checksum, and located at the intersection of the inconsistent checksum row and checksum column. The error is corrected by using the redundancy inherent in the checksum. If the erroneous value is determined to be in the $(i,j)^{th}$ position, and $a_{i,N+1}$ is the ith element of the checksum column, the correct value is given by

$$a_{ij} = a_{i,N+1} - \sum_{\substack{k=1 \\ k \neq j}}^{N} a_{ik}$$

Use of the row weighted checksum matrix was considered by Jou and Abraham [16] and Luk and Park [8]. Jou and Abraham suggest using $w_i = 2^{(i-1)}$. For large matrices, this suffers from the size of the weights. Indeed for floating point arithmetic the weights can become large enough that errors in low numbered rows and columns can become lost in the roundoff even with relatively small matrices. Luk and Park [8] suggest a different weighting with $w_i = i$. These weights are appropriate for the row weighted checksum matrix, but in the natural extension to multiple checksum rows for detecting multiple faults (see the following) they become too large. Brent et al. [17] developed a method to determine small weights for multiple checksum rows. Nair and Abraham [19)] consider other encoding vectors that have numerical advantages when there is *a priori* knowledge about the entries in the matrix (such as sign biases), but the weights are not unique, so these encodings can only be used to detect errors, not locate and correct them. With the row weighted checksum matrix, the error is detected and corrected in the same way as with the full checksum matrix. The erroneous value is located by using the fact that

$$(Aw)_i - <a_i,w> = w_j ((Ae)i - <a_i,e>)$$

if the error is in row i and column j, where $(Ax)_i$ is the ith element of a checksum column, a_i is the ith row of the information matrix, and $<\bullet, \bullet>$ is the usual l_2 inner product. In other words, if the weighting $w_i > = i$ is used, the ratio of the errors in the checksum columns gives the column in which the error is located.

Huang and Abraham [1] show that there are at least five matrix operations that preserve the checksum property.

1. Addition

2. Multiplication

3. Scalar product

4. LU decomposition

5. Transpose

For example, consider two matrices, A and B, and their product $C = AB$. The product of the column checksum matrix for A and the row checksum matrix for B is the full checksum matrix for C as follows.

$$\begin{bmatrix} A \\ e^T A \end{bmatrix} [B \ Be] = \begin{bmatrix} AB & ABe \\ e^T AB & e^T ABe \end{bmatrix} = \begin{bmatrix} C & Ce \\ e^T C & e^T Ce \end{bmatrix}$$

With the matrix encodings of the previous discussion, it is possible to correct a single error. However, it is easily seen that in most cases these encodings are capable of detecting many errors, and in special situations they can be used to correct more than one error.

With the full checksum matrix, one can detect one error in each row (or each column), and could in fact correct those errors if they could be located. However, there is insufficient information to locate them, except for the special case when all the errors are in the same column (or same row), assuming that the errors do not cancel each other in the checksum. If all the errors are in the same row, there is one inconsistent checksum in the checksum column, indicating the row of the errors, and several inconsistencies in the checksum row, indicating the columns of the errors. The checksum row is used to correct each of the errors.

With the row weighted checksum matrix, the detection and location of an error is independent for each row. Thus, this approach allows the detection, location and correction of up to N errors, provided each error is in a different row.

Both methods have the ability to correct up to N errors provided the errors lie in the right configuration. To further determine which method can handle more cases, one must examine the number of error configurations that each method can handle in an $N \times N$ matrix. By an *error configuration* is meant the specification of a set of locations in the matrix where the values are erroneous.

For the full checksum matrix, note that there are $2^N - 1 - N$ ways to choose between 2 and N locations for errors in one row. There are N rows, so there are $N(2^N{-}1{-}N)$ configurations that contain more than one error with all the errors in the same row. There are also the same number of configurations with all the errors in the same column. Finally, including the N^2 configurations with one error, there are $N2^{(N+1)} - 2N - N^2$ error configurations where the full checksum matrix provides enough information to correct all the errors.

For the row weighted checksum matrix, consider the number of configurations with exactly k errors, none of which lie in the same row. There are N^k ways to choose the k rows that contain an error. Each row can have its error in any of the N locations in the row, hence there are N^k ways to choose the locations in the rows. The total number of error configurations that contain between 1 and N errors with no two errors in the same row is then

$$\sum_{k=1}^{N} N^k \binom{N}{k} = (N+1)^N - 1$$

One should also consider what the chances are that the methods would fail to detect the presence of errors. This can happen in the full checksum matrix approach only when both there is exactly one row (or column) in which the errors sum to a nonzero value and there is at least one row (or column) in which the errors sum to exactly zero (cancel out). The row weighted checksum matrix approach can fail to detect an error when there is more than one error in one row. Since the error in a row is located by the ratio of the values in the two checksum columns, if there is more than one error in a row and this ratio is exactly an integer between 1 and N, then the method will "locate" a single error in the wrong column, "correct" a possibly already correct value to an incorrect value and report the result as correct. In terms of the chances of these situations arising, it is easily shown that for both approaches, the described situation occurs with "probability 0" in a measure theoretic sense. However, when working with finite precision on a digital computer, there is a small but nonzero probability of these situations arising, but it is small enough to be acceptable for either of the approaches.

By using more than two checksum columns and/or checksum rows it is possible to detect, locate and correct more errors. Jou and Abraham [16], Park [10], and Brent et al. [17] propose multiple weighted checksum columns to manipulate multiple errors. The multiple row weighted checksum matrix is give by

$$[A \; AW]$$

where W is a matrix of weights. Jou and Abraham propose that $w_{ij} = 2^{(i-1)(j-1)}$, Park proposes $w_{ij} = i^{(j-1)}$, and Brent et al. provide a scheme to determine weights that do not grow exponentially with either N or the number encoding vectors. Park shows that, for matrix triangulation, all the errors in up to t columns can be detected using t weighted checksum columns. However, the errors cannot in general be located and corrected. Anfinson and Luk [18] show how to correct two errors with four checksum columns.

A second approach extending the checksum techniques to multiple faults was briefly mentioned by Huang and Abraham [1], and considered by Park [10]. In the *block checksum scheme* the matrix is partitioned into p submatrices, and each submatrix is encoded as a full or row weighted checksum matrix. Each checksum submatrix is encoded as a full or row weighted checksum matrix. Each checksum submatrix has the same fault tolerant capabilities as a checksum matrix the same size. This approach can be used with any of the matrix operations for which the single-fault checksum techniques can be used, including matrix multiplication [1] and LU decomposition [10]. The hardware overhead is the same as that invoked by the multiple weighted checksum columns approach, whereas the preceding computation is as less a result of the shorter vectors being summed.

The Connection Machine is a multiprocessor with tens of thousands of processors, usually 16 K or 64 K where K = 1024. By using the NEWS networks (North, East, West, South), the Connection Machine can be considered to be a mesh connected array. Software is provided for "virtual processor sets" that allow the user to program the machine as if it had more processors than are physically available. The Connection machine is thus very well suited to matrix operations by configuring the machine as a two-dimensional NEWS network of virtual processors with one virtual processor per matrix element.

In the Connection Machine software there is a restriction on the number of rows and number of columns in the virtual processor set—they must be powers of 2. This restriction motivates an approach to algorithm-based fault tolerance on the Connection Machine. When performing operations on matrices whose sizes are not powers of 2, there will be some rows and some columns of virtual processors that are idle. These rows and columns can be used for the checksum vectors with virtually no hardware overhead. Thus the number of checksum vectors in our fault-tolerance scheme is determined by examining the size of the matrix relative to powers of two. In most cases, for N rows (or columns) the number of checksum vectors used will be $2^k - N$ where k is such that $2^{k-1} \le N < 2^k$. One may choose to increase to the next virtual processor set size when N is less than, but very close to, a power of 2, to provide a higher degree of multiple fault handling. This provides the basic scheme for ABFT on the Connection Machine. What remains is to determine the most appropriate approaches.

For the choice between using multiple weighted checksum columns to encode the information matrix and encoding the submatrices of a partitioning of the information matrix, the latter is more appropriate for the Connection Machine. Three considerations led to this conclusion. With the exception of the weights proposed by Brent et al. [17], the weights quickly dominate the number of significant digits in floating point numbers for the large matrices manipulated in the Connection Machine. The weights recently proposed by Brent et al. may alleviate this problem, but they do not affect the remaining two considerations. If only checksum columns were used (no checksum rows), the excess virtual processors in the row dimension would remain idle. Thus by partitioning in both directions a higher processors utilization can be obtained. Finally, it is easily seen that the overhead in computing the checksum vectors is smaller for the partitioned matrix approach. This is because there is no multiplication by weights, and the summations are shorter and summation of N numbers requires $O(LOGN)$ time.

To encode the submatrices of the partitioned information matrix, there is a choice between the full checksum matrix approach and the row weighted checksum approach. The full checksum matrix approach is more appropriate for the Connection Machine. As noted, if only checksum columns are used there will be idle virtual processors that could be used to assist the error checking process. More important, the full checksum matrix uses only one checksum column per submatrix, whereas the row weighted checksum matrix uses two checksum columns per submatrix. With a fixed number of checksum columns available, as determined by the number of columns in the information matrix and the number of columns in the virtual processor set, the information matrix will have twice as many partitions, that is, the submatrices for the full checksum matrix approach have half as many columns as the submatrices for the row weighted checksum matrix approach. We again see the smaller overhead of computing the checksum columns for the same reasons as the previous paragraph.

Also in favor of the full checksum matrix approach is the number of error configurations that can be handled. The row weighted checksum matrix can handle more configurations of erroneous values than the full weighted checksum matrix. However, this assumes that the matrices being encoded by the two approaches are of the same size. It is now seen that a more appropriate comparison is obtained by assuming the full checksum is encoding a matrix of half the size in each dimension. Thus, if the submatrix encoded by the row weighted checksum matrix is $N \times N$, the same submatrix is actually four independent submatrices of size $N/2 \times N/2$ for the full checksum encoding. Recalling that the full checksum matrix can handle $N2^{N+1} - 2N - N^2$ error configurations for an $N \times N$ matrix, the number of error configurations that can be handled in four independent matrices of size $N/2 \times N/2$ is

$$\left[\frac{N}{2} 2^{\frac{N}{2}+1} - 2\frac{N}{2} - \left(\frac{N}{2}\right)^2 \right]^4$$

$$= (N2^{N/2} - N - N^2/4)^4$$

It is easily seen that the number of error configurations the row weighted checksum matrix can handle, $(N + 1)^N - 1$, is asymptotically larger, but is smaller for small values of N. The crossover point occurs at about $N = 9$. By examining the size of the submatrices obtained by the scheme described previously as a function of the number of rows and columns in the information matrix, it can be seen that the submatrix sizes are less than 9 about two-thirds of the time. Therefore, the full checksum matrix approach can handle more error configurations than the row weighted checksum approach when using the partitioning scheme described in the preceding.

To summarize, the checksum scheme proposed for the Connection Machine is to use the full checksum matrix encoding of the submatrices obtained by partitioning the information matrix into enough parts to use all the idle processors (of the virtual processor set dictated by the size of the information matrix) as checksum elements. In the case where the dimension of the information matrix is exactly a power of 2, the next sized virtual processor set must be used. The following algorithm shows how to create such a checksum matrix on the Connection Machine from an $N \times M$ matrix A residing on the front end computer (host).

1. Determine c_1 such that $2^{c_1-1} \leq N < 2^{c_1}$ and c_2 such that $2^{c_1-1} \leq M < 2^{c_2}$.

2. Let $d_1 = 2^{c_1}$ and $d_2 = 2^{c_2}$.

3. Let the number of checksum rows be $d_1 - N$ and checksum columns be $d_2 - M$.

4. Determine the partitioning so as to create nearly equal sized submatrices with the number of partitions determined by the number of checksum columns and rows.

5. Copy A to the first N rows and M columns of a $d_1 \times d_2$ virtual processor set.

6. Spread A over the virtual processor set to provide an empty column or row between each partition.

7. Use the Connection Machine "scan" operation to sum the rows of the submatrices and another scan operation to sum the columns of the submatrices.

The next algorithm shows how to detect, locate, and correct any errors in the result checksum matrix.

1. Compute checksums using the "scan" operation.

2. Compare computed checksums with the result's checkvectors and set an inconsistency flag if they disagree.

3. Determine the number of inconsistencies in each checkvector.

4. Set error configuration type (ECT) to be:

   0 if no inconsistencies

   1 if 1 inconsistency in checkrow and 1 inconsistency in checkcolumn

   1 if > 1 inconsistency in checkrow and 1 inconsistency in checkcolumn

   2 if 1 inconsistency in checkrow and >1 inconsistency in checkcolumn

   3 otherwise (type 3 errors are detected but cannot be corrected)

5. Pass checkcolumn inconsistency flag across rows and checkrow inconsistency flag up columns, and "and" them to locate errors.

6. Compute correction and correct error using the checkrow where ECT $= 2$.

Algorithm-based fault tolerance provides the possibility of fault tolerance with a very small overhead, perhaps on the order of 5 to 10% as opposed to the 300% or more required by traditional approaches. Huang and Abraham [1], for example, show that the overhead for matrix multiplication using the full checksum matrix is $O(1/N)$ in hardware and $O(\log(N)/N)$ in time. It is easily seen that the amount of overhead decreases with N, and even for moderate sized N the overhead is small. However, an analysis of this type considers only the hardware and floating point operations required to compute the checksums and check for consistency. In practical implementations, there will be other overheads that are difficult to compare to floating point operations. For example, in the approach presented one must determine how to partition the matrix, and then spread the matrix over the processor set leaving empty rows and columns for the checksums.

Here, the results of numerical experiments to determine the time overhead of an implementation on a Connection Machine, using matrix addition and matrix multiplication as examples are presented. Since this approach uses existing processors that would otherwise be idle, hardware overhead is nonexistent, although this assumption might be debatable.

These computations were performed on a Thinking Machines CM-2 with 16 K processors, although only 8 K processors were used, and without floating point hardware. The front-end computer (host) was a VAX 8350 operating under ULTRIX V2.3. Programs were written in C calling PARIS subroutines for the Connection Machine operations, and were compiled with the ULTRIX C compiler.

The overheads are divided into four parts, each associated with some operation of a non-fault-tolerant program, as follows.

Normal Operation	Overhead
Create geometry, VPset, and fields for matrices	Determine partition, allocate fields for partition flags, and set partition flags
Load matrices from host to CM	Spread matrices over partition and compute checksums
Matrix operation (add or multiply)	Overhead within operation itself and detect, locate, and correct errors
Return result from CM to host	Collapse matrix from partition

Each of these operations can occur one or more times in a program, depending on the application. The overhead associated with the matrix operation is further broken into two parts: that associated with the matrix operation itself, and that owing to the detection, location, and correction of errors. To obtain a more general result, each of the five overheads was measured individually. The total of the five measurements is equivalent to a program that performs the matrix operation once. The CPU time was measured using the PARIS routines CM_reset_timer, CM_start_timer and CM_stop_timer. To minimize the effect of timing variability, the measured times were averaged over 20 runs, and the computations were performed during off hours on a dedicated machine. Measurements were made for $N \times N$ matrices with all values of N from 4 to 32, and 20 exponentially distributed values of N between each power of 2 from 32 to 512, that is, N of the form $(2^{i/20})$ where i is an integer between 100 and 180.

Examination of the results for matrix multiplication reveals many interesting observations about the behavior of the Connection Machine in general as well as the fault-tolerant algorithm. Most of the behaviors can be easily explained, but a few remain a mystery. The NFT (nonfault tolerant) set up time is essentially constant, as one would expect. The FT (fault tolerant) set up time is constant except for jumps at $N = 128$ and $N = 256$ where the *VP ratio* (the ratio of the number of virtual processors to the number of physical processors) increases by a factor of 4. This increase is expected since each virtual processor must determine whether it is to contain a checksum value or a value from the information matrix. Since a physical processor must perform the operations of all the virtual processors to which it is assigned, one would expect to see such a jump when the VP ratio increases.

For large N the time to load matrices into the Connection Machine (from the front-end) increases as $O(N^2)$ data elements to transfer. There is a slight jump when the VP ratio increases. For small N, the time is dominated by low order effects, probably involving a "start up" overhead. An unexplained feature is the increase in time between $N = 64$ and $N = 90$. It is suspected that this may be related to the fact that 8 K processors are being used, since the range corresponds exactly to when the VP ratio is larger than 1, but N^2 is less than 8 K. The time for returning the results to the front-end is similar, but only half as large because two matrices are loaded and one is returned.

The time for NFT matrix multiplication increases linearly with N, except for jumps when the VP ratio increases, as one would expect for an $O(N^3)$ operation algorithm on N^2 processors. The time for FT matrix multiplication is a step function. This is because the size of the checksum matrix is always increased to the next power of 2 by adding checkvectors, so the matrix size is constant between powers of 2. The time to detect, locate, and correct the errors is constant except for jumps when the VP ratio increases. It is not known why this is the only operation for which there is a jump at $N = 64$, the first time the VP ratio increases.

The total time is dominated by the transfer of data between the front-end and Connection Machine for small N. The set up time and the time to detect, locate, and correct errors are small compared to the other times.

The overhead invoked by fault-tolerance for matrix multiplication is given by

$$\text{Overhead} = \left(\frac{\text{time for FT version}}{\text{time for NFT version}} - 1 \right) \times 100\%$$

The set up overhead is extremely large. However, the time for set up is very small, and this has virtually no effect on the total overhead.

For small N, the overheads associated with transferring data between the front-end and Connection Machine (load matrices and return result) decrease from less than 200% to considerably less than 100%. Loading has a larger overhead because this includes computing the checksum vectors. For large N, the overhead decreases between powers of 2 with a jump increase at powers of 2. The overhead is larger when N is slightly larger than a power of 2 because the values from the information matrix must be moved farther when there are more checkvectors. There is a small spike at powers of 2 where the VP ratio increases earlier for FT than for NFT (see the following). Generally, overhead for large N is between 40 and 15% for loading, and between 27 and 10% for returning.

The overhead for the matrix multiplication generally decreases from 100% at powers of 2 to near 0% when N is slightly less than a power of 2. This is expected since the matrix size is increased to the next power of 2, and hence is $2N$ when N is a power of 2 and $N + 1$ when N is one less than a power of 2. The most prominent features of the matrix multiplication graph are the large spikes at $N = 128$ and $N = 256$. For large N that are powers of 2, the overhead can be as large as 700% because of the combination of two effects. First, as noted, the size of the matrices being multiplied are $2N$, requiring twice the time of the NFT version. Second, the VP ratio is increased by a factor of 4 at a power of 2 for FT, and at 1 plus a power of 2 for NFT. Hence, at the power of 2, FT can take up to eight times as long as NFT.

Next consider the overhead for the combined operations of matrix multiplication and detecting, locating and correcting the errors. For small N, the overhead decreases in a sawtooth manner from about 250 to about 10%. For large N, the overhead is dominated by the matrix multiplication, and the cost of fixing errors is negligible.

The overhead for the total is similarly dominated by different operations for different values of N. For small N, the overhead is dominated by those associated with the data transfer operations and detecting, locating and correcting the errors. It begins at 200% and decreases until the overhead involved with matrix multiplication dominates. For large N, the overhead is similar to the overhead associated with the matrix multiplication itself.

For N sufficiently large, the overhead can be characterized as follows. If N is a power of 2, the overhead is unacceptably large due to the combined effects of doubling the matrix size and quadrupling the VP ratio. If N is slightly larger than a power of 2, the overhead is near 100%, whereas if N is slightly smaller than a power of 2, the overhead is near 0%. However, the large overhead for N slightly larger than a power of 2 is related to a higher degree of fault tolerance. The overhead is owing to the large number of checkvectors, which enable the algorithm to correct a larger number of errors. In practice, one would modify the algorithm such that the number of checkvectors is selected according to an expected number of simultaneous faults. Unless one expects many processors to fail, the overhead would be near 0% for all large N except powers of 2, and possibly N slightly less than powers of 2 if multiple fault capability is required.

In assessing overhead the results for matrix addition only the total time is considered. The overhead for set up, host/CM data transfers, and fixing errors are identical to those from matrix multiplication. The time for the matrix addition itself is piecewise constant (jumps occur when the VP ratio increases), and is about 1 millisecond for small N. There is no difference between the fault-tolerant and non-fault-tolerant matrix additions.

The matrix addition operation is dominated by the transfer of data between the host and Connection Machine. For small N, the detection, location, and correction of errors also makes a noticeable contribution to the total time. The overhead for small N begins at slightly more than 200% and decreases to about 100%.

For large N, the overhead is about 100% at powers of 2, and decreases linearly to about 40% when N is slightly less than the next power of 2.

9.4.1   Summary

A reasonable choice of technique for algorithm-based fault tolerance on the Connection Machine is to use the full checksum encoding of the submatrices of a partitioning of the information matrix, where the partitioning is determined by the number of unused processors in the virtual processor set. We showed how these methods can be implemented on the Connection Machine. Finally, we presented numerical results to illustrate the amount of overhead actually invoked through the use of algorithm-based fault tolerance on the Connection Machine. For large matrices, the overhead is: (1) near 0% when size of the matrix, N, is slightly less than a power of 2; (2) about 100% when N is slightly larger than a power of 2; and (3) as large as 700% when N is exactly a power of 2. The first case is very desirable. The second case coincides with a high degree of fault tolerance, and can be reduced to near 0% at the expense of reducing the number of simultaneous faults handled. The third case is a combination of a high degree of fault tolerance and a premature increase in the virtual processor ratio. This overhead can be reduced in the same way as the second case, but unfortunately the increased virtual processor ratio will keep the overhead at about 400% when N is exactly a power of 2.

REFERENCES

1. Huang, K. and Abraham, J.A., Algorithm-based fault tolerance for matrix operations. *IEEE Trans. Comp.* Vol. C-33, pp. 518–528, 1984.

2. Abraham, J.A., Fault tolerance techniques for highly parallel signal processing architectures. *SPIE Vol. 614 Highly Parallel Signal Processing Architectures*, pp. 49–65, 1986.

3. Choi, Y. and Malek, M., A fault-tolerant FFT processor. *Proc. 15th Int. Symp. Fault-Tolerant Comp.*, pp. 266–271, 1985.

4. Jou, J. and Abraham, J.A., Fault-tolerant FFT networks. *Proc. 15th Int. Symp. Fault Tolerant Comp.*, pp. 338–343, 1985.

5. Jou, J. and Abraham, J.A., Fault-tolerant matrix arithmetic and signal processing on highly concurrent computing structures. *Proc. IEEE,* Vol. 74, pp. 732–741, 1986.

6. Redinbo, G.R., Fault-tolerant convolution using real systematic cyclic codes. *Proc. 17th Int. Symp. Fault-Tolerant Comp.*, pp. 210–215, 1987.

7. Huang, K. and Abraham, J.A., Low cost schemes for fault tolerance in matrix operations with processor arrays. *Proc. 12th Int. Symp. on Fault-Tolerant Computing*, pp. 330–337, 1982.

8. Luk, F.T. and Park, H., An analysis of algorithm-based fault tolerance techniques. *J. Par. and Dist. Comp.,* Vol. 5, pp. 172–184, 1988.

9. Luk, F.T. and Park, H., Fault-tolerant matrix triangularization on systolic arrays. *IEEE Trans. Comp.,* Vol. 37, pp. 1434–1438, 1988.

10. Park, H., Multiple error algorithm-based fault tolerance for matrix triangularizations. Tech. report TR88-73, Computer Science Dept., Univ. of Minn., Minneapolis, MN, 1988.

11. Luk, F.T., Torng, E.K., and Anfinson, C.J., A novel fault tolerance technique for recursive least squares minimization. *J. VLSI Signal Processing,* to appear.

12. Aykanat, C. and Ozguner, F., A concurrent error detecting conjugate gradient algorithm on a hypercube multiprocessor. *Proc. 17th Int. Symp. Fault-Tolerant Comp.,* pp. 204–209, 1987.

13. Huang, K. and Abraham, J.A., Fault-tolerant algorithms and their application to solving Laplace equations. *Proc. 1984 Int. Conf. on Parallel Processing,* pp. 117–122, 1984.

14. Mitchell, William F., *Algorithm-Based Fault Tolerance for Matrix Operations on the Connection Machine,* Advanced Technology Laboratories, General Electric Company, Moorestown, NJ; Technical Report CMAT-89-TR-003.

15. Hamming, R.W., Error detecting and error correcting codes. *Bell System Tech. J.,* Vol. 29, pp. 147–160, 1950.

16. Jou, J. and Abraham, J.A., Fault-tolerant matrix operations on multiple processor systems using weighted checksums. In Bromley, K., ed., *Real Time Signal Processing VII Proc. SPIE,* Vol. 495, pp. 94–101, 1984.

17. Brent, R.P., Luk, F.T., and Anfinson, C.J., Choosing small weights for multiple error detection. *High Speed Computing II Proc. SPIE,* Vol. 1058, 1989.

18. Anfinson, C.J. and Luk, F.T., A linear algebraic model of algorithm-based fault tolerance. *IEEE Trans. Comp.* Vol. 37, pp. 1599–1604, 1988.

19. Nair, V.S.S. and Abraham, J.A., General linear codes for fault-tolerant matrix operations on processor arrays. *Proc. 18th Int. Symp. Fault-Tolerant Comp.,* pp. 180–185, 1988.

9.5   KALMAN FILTERS

The Kalman filter is an optimal recursive estimator for linear dynamic systems with noise [1]. Many digital signal processing applications require the real time computation of the Kalman filter, for example, tracking an aircraft with radar.

Consequently, many papers have been published on the use of parallel processing technology in the form of systolic arrays for fast computation of the Kalman filter [2–8]. Some of these implementations [4, 7, 8] take the approach of connecting a series of systolic arrays, each of which computes one of the equations of the Kalman filter. Unfortunately, the Kalman filter cannot be pipelined since the result of the last equation is needed as input for the first equation. This results in an excess of hardware with many processors idle at any given time. In another approach [3], the equations are reformulated such that eight applications of the Faddeev algorithm produce the desired result. Thus, the systolic array is programmed for the Faddeev algorithm and the data flow through the array eight times during each iteration of the Kalman filter. However, the Faddeev algorithm requires doubling the dimension of the matrices, and consequently the systolic array takes more processing elements and time steps than necessary. This section [11] presents a new systolic array implementation of the Kalman filter that is not excessive in either hardware or computation steps. For a dynamic system with N states and M observations components, the array uses $N(N + 1)$ processors and about $4N + 6M$ computation steps.

Use of the Kalman filter assumes the dynamic system has a state model of the form

$$x(k + 1) = Ax(k) + w(k) \tag{1a}$$

$$y(k) = Cx(k) + v(k) \tag{1b}$$

where
$x(k) = N \times 1$ state vector at time k
$y(k) = M \times 1$ measurement vector
$A = N \times N$ state transition matrix
$C = M \times N$ measurement transition matrix
$w(k) = N \times 1$ process noise vector
$v(k) = M \times 1$ measurement noise vector
$N =$ number of states
$M =$ number of measurements

It is assumed that $w(k)$ and $v(k)$ are zero mean Gaussian random variables.
The Kalman filter is defined by the following equations.

$$\hat{x}(k|k - 1) = A\hat{x}(k - 1|k - 1) \tag{2a}$$

$$P(k|k - 1) = AP(k - 1|k - 1|k - 1)A^T + Q(k - 1) \tag{2b}$$

$$K = P(k|k - 1)C^T[CP(k|k - 1)C^T + R(k)]^{-1} \tag{2c}$$

$$\hat{x}(k|k) = \hat{x}(k|k - 1) + K[y(k) - C\hat{x}(k|k - 1)] \tag{2d}$$

$$P(k|k) = P(k|k - 1) - KCP(k|k - 1) \tag{2e}$$

where
$$\hat{x}(k|k-1) = \text{the time } k \text{ state estimate given measurements through time}$$
$$k-1$$
$$\hat{x}(k|k) = \text{the time } k \text{ state estimate given measurements through time } k$$
$$P(k|k-1) = \text{the estimate of the state covariance at time } k \text{ given measure-}$$
$$\text{ments through time } k-1$$
$$P(k|k) = \text{the estimate of the state covariance at time } k \text{ given measure-}$$
$$\text{ments through time } k$$
$$Q(k) = E[ww^T] = \text{the process noise covariance matrix}$$
$$R(k) = E[vv^T] = \text{the measurement noise covariance matrix}$$
$$K = \text{the Kalman gain matrix}$$
$$E[\ ] = \text{the expectation operator}$$

The Kalman filter defined by equation 2 can be applied to any dynamic system satisfying the assumption that its state model has the form of equation 1.

Each iteration of the Kalman filter takes a new observation vector $y(k)$, and produces an estimate of the system state at time k, $x(k|k)$. This is a computationally intensive ($O(N^3)$ operations) process involving several matrix multiplications and some form of matrix inversion (or multiplication by the inverse of a matrix). Parallel processing in the form of systolic arrays is known to be very effective for numerical algorithms of this type, using $O(N)$ time and $O(N^2)$ processors.

A systolic array is a parallel processor consisting of several simple processing elements (PEs) connected as a linear array, rectangular mesh, or some other suitable arrangement. Conceptually, the size of the array is selected to match the size of the problem. In practice, one processor may be used to simulate several processors. Generally, data flow in one or two sides of the array, and the results flow out another side.

Systolic arrays are very efficient at computing certain numerical algorithms. For example, there are two well known systolic approaches to computing the matrix product-sum $D = AB + C$, where A, B, C, and D are matrices of appropriate dimensions. If A, B, C, and D were all $N \times N$ matrices, both algorithms use N^2 processors arranged as an $N \times N$ square array, and compute the product in $3N$ time steps, where a time step is the time required for a scalar multiplication and a scalar addition. The first results are complete and available for use after N time steps. Thus, an algorithm that can pipeline several matrix product-sums uses essentially N time steps for each product-sum. The *stationary result matrix multiplication* algorithm begins with C in the array, one element per processor as illustrated in Figure 9.5-1. The matrix A is passed into the left side of the array, and the matrix B into the top. The result, D, resides in the array when the algorithm completes. The other approach, *moving result matrix multiplication*, is illustrated in Figure 9.5-2. B is initially in the array, A is passed to the left, C is passed to the top, and D flows out the bottom. These two systolic approaches to matrix multiplication, and their complementary stationary result-moving result properties, are instrumental in the systolic Kalman filter described in the following.

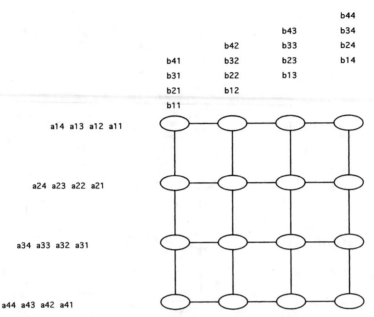

Figure 9.5-1.   Stationary result matrix multiplication of two N × N (N = 4) matrices A and B with addition of an N × N matrix C. The result ends up in the systolic array.

If the notation [A|B] denotes a matrix in which the first columns are those of A and the last columns are those of B, and the k's and hats are dropped for notational convenience, Equations (2a)–(2e) can be rewritten in the following form.

$$T_1 \leftarrow PA^T \tag{3a}$$

$$[x|P] \leftarrow A[x|T_1] + [0|Q] \tag{3b}$$

$$T_2 \leftarrow PC^T \tag{3c}$$

$$[t_1|T_3] \leftarrow C[-x|T_2] + [y|R] \tag{3d}$$

$$\hat{Q}\hat{R} \leftarrow T_3 \tag{3e}$$

$$[t_2|T_4] \leftarrow \hat{Q}^T[t_1|T_2{}^T] \tag{3f}$$

$$[t_3|T_5] \leftarrow \hat{R}^{-1}[t_2|T_4] \tag{3g}$$

$$[x|P] \leftarrow T_2[t_3|-T_5] + [x|P] \tag{3h}$$

For multiplying by the inverse of a matrix (part of the matrix K in Equations (2c)–(2e)), QR factorization is used. In Equation (3e), the matrix T_3 $(= CPC^T + R)$ is factored into an orthogonal matrix Q and an upper triangular matrix R. Multiplication by T_3^{-1} is accomplished in Equations (3f) and (3g) by multiplying by

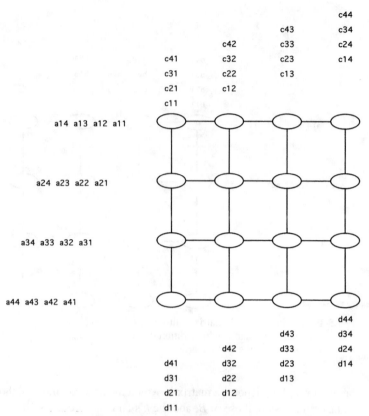

Figure 9.5-2. Moving result matrix multiplication of two N × N (N = 4) matrices A and B with addition of an N × N matrix C. The result D flows out the bottom.

$Q^T (= Q^{-1}$ because Q is orthogonal) and by R^{-1} (easily performed by backward substitution).

Note that, except for Equation (3e), these equations take the form $D = AB + C$,

where
C = sometimes 0

Moreover, the result, D, of one equation is either A or B of the next equation, including wrapping around from Equations (3h) to (3a). This suggests using a rectangular systolic array and alternating between the stationary result and moving result algorithms, cycling through the array eight times to compute the eight equations. This approach works very well, with the timing such that results of one step become available exactly when the next step is ready to begin, providing perfect pipelining of the eight equations, including the wraparound from Equations (3h) to (3a). For clarity, the following considers only the matrix portions of Equations (3a)–(3h), not the augmented vector part (x, y, etc.).

In the first step, Equation (3a) is computed. P is passed in the left side of the array, A^T in the top, and a stationary result matrix multiplication algorithm is executed. After N time steps, the first entry of T_1 is completed, and the first processor is ready to begin the second step, computing Equation (3b). In this step, A is passed to the left, Q in the top, and a moving result matrix multiplication algorithm is executed. After a total of $2N$ time steps, the result $P(k|k-1) = AP(k-1|k-1)A^T$ begins to flow out the bottom of the array. Also at this time, the first processor completes its computations for Equation (3b). As $P(k|k-1)$ flows out the bottom of the array, it is fed into the left side to begin computation of Equation (3c). The third and fourth steps, computing Equations (3c)–(3d), are the basically the same as the first two steps. Note the timing allows perfect pipelining of the equations, and that, since the matrix multiplication algorithms require $3N$ time steps to complete, there are three equations being computed simultaneously in different parts of the array after the computation of Equation (3c) begins.

As T_3 begins to flow out the bottom of the array, it is fed into the left side of the array to begin the QR decomposition algorithm to compute Equation (3e). Givens rotations are used for this decomposition in a triangular subarray as illustrated in Figure 9.5-3. This is a well known systolic algorithm (see, e.g., [9]). The result leaves $\hat{R}$ in the triangular subarray, and the rotation parameters C and S flow out the bottom. These are immediately passed into the left side of the array, while T_2^T is passed in the top, to begin the sixth step. Given the rotation parameters, multiplication by $\hat{Q}^T$ is accomplished by the same algorithm as the "interior" processors of the QR decomposition algorithm. T_4 is left in the array. Meanwhile, the triangular subarray passes the upper triangular matrix $\hat{R}$ out the bottom. $\hat{R}$ is passed in the left side of the array for the back substitution algorithm to multiply T_4 by $\hat{R}^{-1}$. Finally, as values of T_5 become available in the array, T_2 is passed in the left and P in the top for a moving result matrix multiplication for Equation (3h). As the new values of P flow out the bottom, they are immediately passed into the left side of the array to begin the first step of the next iteration.

The Kalman filter systolic array is an $N \times N(N+1)$ rectangular array of processing elements (PEs) each connected to the four nearest neighbors with a two-word data channel, as illustrated in Figure 9.5-4. In this figure, circles represent processors and rectangles represent memory modules. The partitioning of the PEs into eight classes will be explained later. The boundary PEs are connected to memory modules shared with other PEs. Each PE has its own program and a few data registers. Synchronization among the PEs is accomplished by passing data with waits until the data are available from a neighbor or until a neighbor has accepted the previously sent data.

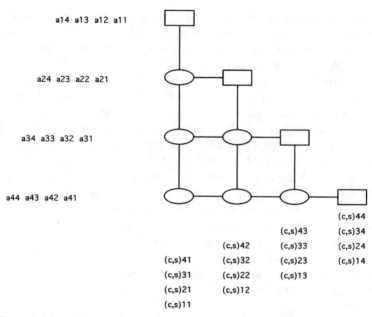

a14 a13 a12 a11

a24 a23 a22 a21

a34 a33 a32 a31

a44 a43 a42 a41

			(c,s)44
		(c,s)43	(c,s)34
	(c,s)42	(c,s)33	(c,s)24
(c,s)41	(c,s)32	(c,s)23	(c,s)14
(c,s)31	(c,s)22	(c,s)13	
(c,s)21	(c,s)12		
(c,s)11			

Figure 9.5-3.   QR decomposition of an $N \times N$ ($N = 4$) matrix A in a triangular systolic array. The transpose of the upper triangular matrix R ends up in the systolic array; the rotation parameters, C and S, of Q flow out the bottom.

The PE programs are based on eight fundamental systolic algorithms. Here, the functionality of the algorithms is briefly described. The algorithms themselves are presented at the end. Note that M, N, A, and so on are generic variables and are not necessarily the same as the number of measurements, number of states, state transition matrix, and so on.

1. **Pass down or Pass right:** These algorithms are used when a PE is not to perform any operations on the data, merely pass it on so that is will eventually leave the array.

2. **Load down:** This algorithm is used to load a vector into a column of PEs for later processing. The top PE receives the first datum, second PE receives the second datum, and so on.

3. **Empty right or Empty down:** These algorithms are used to remove data resident in the systolic array. The data in the left column (or top row) of processors are the first to leave the array.

4. **SRMM:** This is the "stationary result" version of matrix multiplication. The algorithm computes the product of an $N \times L$ array A with an $L \times M$

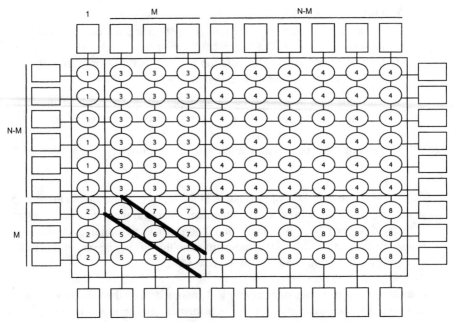

Figure 9.5-4.   Nx(N + 1) rectangular array for the Kalman filter, and partitioning of PEs into eight classes.

array B in an $N \times M$ systolic array. The rows of A are passed in the left side of the systolic array and the columns of B are passed in the top. The product ends up in the array.

5. **MRMM:** This is the "moving result" version of matrix multiplication. The algorithm computes the product and sum $AB + C$ where A is an $N \times L$ array, B is an $L \times M$ array and C is an $N \times M$ array. It uses an $L \times M$ systolic array. B is resident in the array registers initially. The columns of A are passed in the left side of the array and the columns of C are passed in the top. The result comes out the bottom of the array.

6. **QR(int) and QR(diag):** These algorithms work together to perform QR decomposition. The algorithms decomposes an $N \times M$ matrix A into the product of an $N \times M$ orthogonal matrix Q and $M \times M$ upper triangular matrix R by using Given's rotations in an $M \times M$ lower triangular systolic array. The columns of A enter the array from the left. The rotation parameters for Q come out the bottom. R (or actually the transpose of R) ends up in the array. There are two algorithms depending on whether the PE is on the diagonal (top) or is interior (all others).

7. QM: This algorithm multiplies by the transpose of O. It multiplies an $N \times L$ matrix A by the transpose of the $N \times M$ orthogonal matrix O using the rotation parameters (C,S) output from the preceding algorithm. A enters the array from the top; the rotation parameters for O enter from the left. The result is left in the $M \times L$ systolic array.

8. Multiplication by the inverse of an upper triangular matrix (Uinv): This algorithm multiplies an $M \times N$ matrix A by the inverse of an $M \times M$ upper triangular matrix U. A is in the $M \times N$ systolic array initially. The rows of U are passed int the left side of the array. The result is left in the array.

There are eight classes of processors in the array, as illustrated in Figure 9.5-4, with all PEs of the same class running nearly identical programs. Each PE program is described by a sequence of eight steps corresponding to the eight Equations (3a)–(3h), with NULL indicating that this PE has no operations in this step and proceeds immediately with the next step.

Boundary elements read and write directly to memories shared by other boundary elements in lieu of receiving and passing data on the boundary sides. To minimize memory access conflicts, each memory is connected to one to four processors and each matrix is distributed over several memories, for example, each column of the matrix P is in a different memory connected to one of the top, side, and bottom PEs.

The specific operations and data memory access required for each PE class and each step are shown at the end. The figures at the end, together with the fundamental algorithms also at the end, fully describe the operation of the Kalman filter systolic array. The following comments should help to clarify some of the details.

1. In general, vectors, matrix rows, and matrix columns are passed into the array in forward order, that is, with the first value first and the Nth value last. The exception is in Uinv where the end of the row is passed in first and the beginning of the row (diagonal entry) is passed in last.

2. Synchronization of the array is accomplished through the passing of data. Thus a "get" includes a "wait for the data to become available", and a "send" includes a "wait for the previously sent data to be used". On the boundary, reading PEs must wait for updated vector and matrix values to be written by writing PEs. A flag is associated with each memory word indicating whether or not a value is ready to be read. In general, the flag is cleared when the value is read by a PE, and set when the value is written to by a PE. The exceptions where the flag is not cleared are P and x in step 3 and T_2 in step 6, where the values will be used again before they are updated. Also, flags are not used with the constant matrices A, C, O, and R.

3. In step 2, ZERO is a vector of N 0s.

4. Some steps, such as step 1 for PE class 1, involve merging two of the fundamental algorithms. This involves interleaving the lines of the algorithms, and possibly modifying the looping structure. For example, "Load down & Pass right" is

```
for loop = 1 to PE_row-1
    get B from left port
    send B right
endfor
get A from top port and B from left port
register = A
send B right
for loop = PE_row+1 to N
    get A from top port and B from left port
    send A down and B right
endfor
```

5. The matrix P is symmetric, thus the rows of P are the same as the columns of P, and a data memory labeled "P row i" (as in step 1) is the same memory as one labeled "P col i" (as in step 2).

6. The matrices A and C also need to be assessed both by row and by column. These matrices can each be stored in two memories, one by row and one by column, since they are constant and there is no concern about updating two locations.

7. The matrix T_2 needs to be accessed by row in step 6 and by column in step 8. It is written by row in step 4, ready for step 6, and can be copied by a small number of other processors to a second memory, accessible by column, in the amount of time available before step 8.

8. x in step 3 and P in step 8 must be negated before reading and writing. The programs in the boundary PEs are modified to perform this operation.

9. When the algorithm calls for passing a value to the next processor and either (1) the next processor is NULL during this step or (2) this processor is on the boundary and there is no memory write during this step, the algorithm in this processor is modified to remove the "send".

A careful analysis of the flow of data through the Kalman filter systolic array reveals the number of cycles required per iteration. By a cycle, we mean the operations that occur in one processor with one pair of data. For example, in stationary result matrix multiplication one cycle consists of two gets, one multiplication, one add, and two sends. By determining the maximum time required for any one cycle, one can estimate the speed of the systolic array using the cycle counts that follow. Table 9-6 shows the staring and ending cycle number for each of the steps of the algorithm. In steps 6 and 8, the start time is the beginning of the actual computation in PE class 2, not the beginning of passing the data down through class 1, 3, and 4 PEs. The $\max\{0, N - 2M - 2\}$ term arises in step 6 because which datum, the top or left, is the first to arrive at the first class 2 PE depends on the relationship between N and M. It is seen that one complete iteration requires $6N + 6M + 1 + \max\{0, N - 2M - 2\}$ cycles. However, due to the overlap of computations between iterations, the latency between beginnings of iterations is $4N + 6M + 3 + \max\{0, N - 2M - 2\}$. For the example of tracking with radar, $N = 9$ and $M = 3$, so each iteration uses 58 cycles before the next iteration can begin. Numerical simulations of the Kalman filter systolic array confirm these complexity estimates.

9.5.1   Eight Fundamental Systolic Algorithms

The PE programs are based on eight fundamental systolic algorithms described here. In these algorithms, M, N, A, and so on, are generic variables and are not necessarily the same as the number of measurements, number of states, state transition matrix, and so on. Loops for which the upper limit is less than the lower limit ("zero pass loops") are not executed.

TABLE 9-6

STARTING AND ENDING CYCLE NUMBERS FOR EACH STEP		
Step	**Start**	**End**
1	1	$3N - 1$
2	$N + 1$	$4N - 1$
3	$2N + 2$	$4N + M$
4	$3N + 2$	$5N + M$
5	$4N + 3$	$4N + 3M + 1$
6	$4N + M + 4 + $ "max"*	$5N + 3M + 2 + $ "max"
7	$4N + 3M_4 + $ "max"	$5N + 5M + 2 + $ "max"
8	$4N + 5M + 3 + $ "max"	$6N + 6M + 1 + $ "max"
1	$4N + 6M + 4 + $ "max"	$7N + 6M + 2 + $ "max"

*"max" $= \max\{0, N - 2M - 2\}$.

1. Pass data down (or right) (Pass down, Pass right): This algorithm is used when a PE is not to perform any operations on the data, merely pass them on so they will eventually leave the array.

```
for loop = 1 to N
        get A from top (or left) port
        send A down (or right)
endfor
```

2. Load data down (Load down): This algorithm is used to load a vector of length N into a column of N PEs for later processing. The top PE receives the first datum, second PE receives the second datum, and so on.

```
get A from top port
register = A
for loop = 1 to N-PE_row
        get A from top port
        send A down
endfor
```

3. Empty data right (or down) (Empty right, Empty down): This algorithm is used to remove data resident in the systolic array. The data in the left column (top row) of processors are the first to leave the array.

```
for loop = 1 to PE_col-1 (PE_row-1)
        get A from left port (top port)
        send A right (down)
endfor
send register right (down)
```

4. Stationary result matrix multiplication (SRMM): This algorithm computes the product of an $N \times L$ array A with an $L \times M$ array B in an $N \times M$ systolic array. The rows of A are passed to the left side of the systolic array and the columns of b are passed in the top. The product ends up in the array.

```
register = 0
for loop = 1 to L
        get A from left port and B from top port
        register = register + A*B
        send A right and B down
endfor
```

5. Moving result matrix multiplication (MRMM): This algorithm computes the product and sum $AB + C$ where A is an $N \times L$ array, B is an $L \times M$

arrays and C is an $N \times M$ systolic array. B is resident in the array registers initially. The columns of A are passed in the left side of the array and the columns of C are passed in the top. The result comes out the bottom of the array.

```
for loop = 1 to N
        get A from left port and C from top port
        send A right and A *register+C down
    endfor
```

6. QR decomposition (QR(int), QR(diag)): This algorithm decomposes an $N \times M$ matrix A into the product of an $N \times M$ orthogonal matrix Q and an $M \times M$ upper triangular matrix R by using Given's rotation in an $M \times M$ lower triangular systolic array. The columns of A enter the array from the left. The rotation parameters for Q come out the bottom. R (or actually the transpose of R) ends up in the array. There are two algorithms depending on whether the PE is on the diagonal (top) or is interior (all others).

Diagonal PEs:
```
for loop = 1 to N
      get A from left port
      if loop = 1 then
            if A = 0 then
                  C = 1
                  S = 0
                  register = 0
            else
                  C = 0
                  S = sgn(A)
                  register = abs(A)
            endif
      else
            if A = 0 then
                  C = 1
                  S = 0
            else
                  r' = sqrt(register*register + A*A)
                  C = register/r'
                  S = A/r'
                  register = r'
            endif
      endif
      send (C,S) down
    endfor
```

Interior PEs:

```
register = 0
for loop = 1 to N
        get A from left port and (C,S) from top port
        temp = C*A - S*register
        register = S*A + C*register
        send temp right and (C,S) down
endfor
```

7. **Multiplication by the transpose of Q via rotation parameters (QM):** This algorithm multiplies an $N \times L$ matrix A by the transpose of the $N \times M$ orthogonal matrix O using the rotation parameters (C,S) output from the preceding algorithm. A enters the array from the top; the rotation parameters for O enter from the left. The result is left in the $M \times L$ systolic array.

```
register = 0
for loop = 1 to N
        get (C,S) from left port and A from top port
        temp = C*A - S*register
        register = S*A + C*register
        send (C,S) right and temp down
endfor
```

8. **Multiplication by the inverse of an upper triangular matrix (Uinv):** This algorithm multiplies an $M \times N$ matrix A by the inverse of an $M \times M$ upper triangular matrix U. A is in the $M \times N$ systolic array initially. The rows of U are passed in the left side of the array. The result is left in the array.

```
for loop = 1 to M-PE_row
        get U from left port and B from bottom port
        register = register -U*B
        pass U right and B up
endfor
get U from left port
register = register/U
pass U right and register up
```

9.5.2   Eight Algorithmic Steps
in the Systolic Array

The algorithms are broken into eight steps corresponding to the eight Equations (3a)–(3h). The processors are partitioned into eight PE classes; all processor within a class contain essentially the same program. The eight steps are illustrated in Figures 9.5-5 to 9.5-12. The data to be read from or written to memory are indicated in the rectangular boxes of the figures.

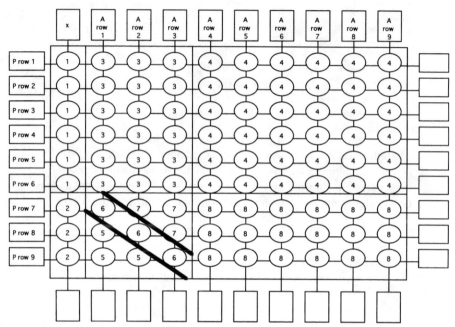

Figure 9.5-5.   Step 1: Operations and data.

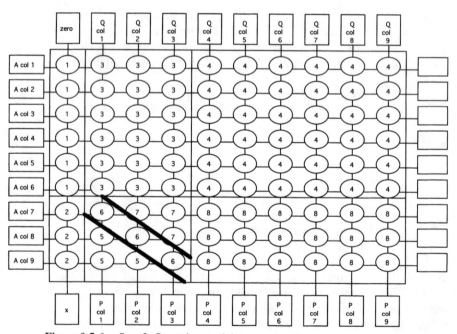

Figure 9.5-6.   Step 2: Operations and data memory access.

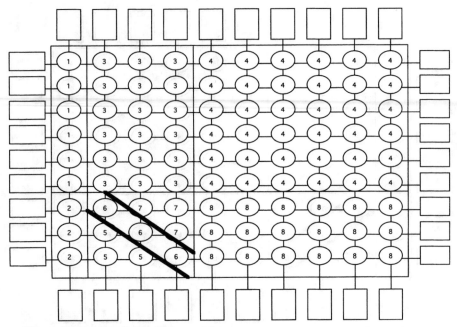

Figure 9.5-7.   Step 3: Operations and data memory access.

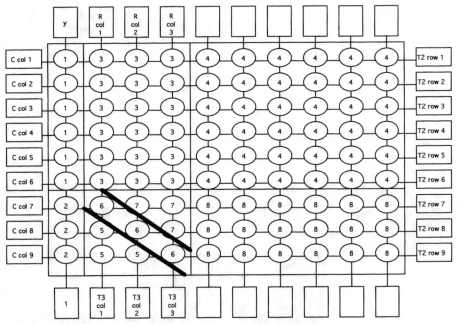

Figure 9.5-8.   Step 4: Operations and data memory access.

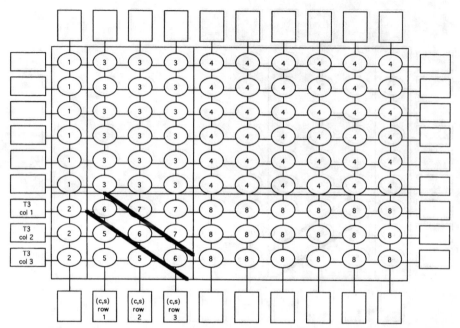

Figure 9.5-9.   Step 5: Operations and data memory access.

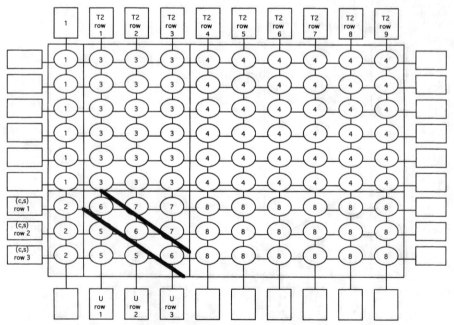

Figure 9.5-10.   Step 6: Operations and data memory access.

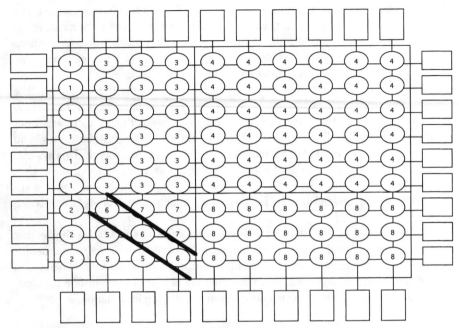

Figure 9.5-11.   Step 7: Operations and data memory access.

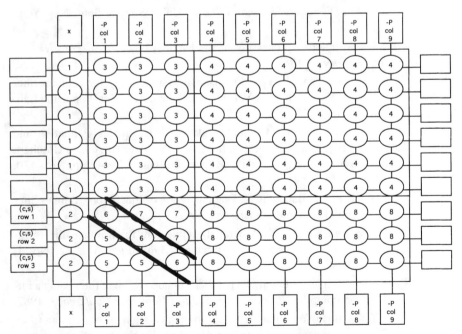

Figure 9.5-12.   Step 8: Operations and data memory access.

A new systolic array implementation of the Kalman filter was presented. The array used $N(N + 1)$ processing elements, connected as a rectangular mesh, fewer than many other proposed Kalman filter systolic arrays. Each processor contains a simple, easily implemented algorithm, and requires only a few data registers. Computation of the Kalman filter equations is pipelined in eight passes through the array, including pipelining of the wraparound from the last equation to the first. The array requires approximately $4N + 6M$ computation steps for each iteration of the filter, fewer than some other approaches. Although there are faster Kalman filter systolic arrays and there are Kalman filter systolic arrays with fewer processors, each of these requires more of the other resource. This systolic array provides a better balance between processors and computation steps. Also, the simplicity of the design makes this approach easy to implement, and this approach is easily extended to be fault tolerant through the use of algorithm-based fault tolerance [10].

The lesson for parallel processing learned from this new algorithm is that new algorithms (and even new implementations of old algorithms) can dramatically affect computational requirements, often in an attractive way. When faced with a shortage of processors, memory, runtime budget or other resource, reconsidering the algorithm can often allow a satisfactory implementation.

References

1. Bozic, S.M., *Digital and Kalman Filtering*. John Wiley & Sons, New York, 1979.

2. Andrews, A., Parallel processing of the Kalman Filter. *IEEE Parallel Processing Conference*, 1981.

3. Chang, J.J. and Yeh, H., Two-dimensional systolic array for Kalman-filter computing. NASA Tech Brief Vol. 12, No. 9, Item #116, Jet Propulsion Laboratory, Pasadena, CA, 1988.

4. Chen, M.J. and Yao, K., On realizations of least-squares estimation and Kalman filtering by systolic arrays. In Moore, W., McCabe, A., and Urquhart, R., eds., *Systolic Arrays*. Adam Hilger, Bellington, WA, pp. 161–170, 1986.

5. Itzkowitz, H.R. and Baheti, R.S., Demonstration of square root Kalman filter on Warp parallel computer. *American Control Conference*, 1989.

6. Jover, J.M. and Kailath, T., A parallel architecture for Kalman filter measurement update and parameter estimation. *Automatica,* Vol. 22, No. 1, pp. 43–57, 1986.

7. Shaffer, P.L., Implementation of a parallel extended Kalman filter using a bit-serial silicon compiler. *Proc. 1987 Fall Joint Computer Conference*, 1987.

8. Sung, T.Y. and Hu, Y.H., Parallel VLSI implementation of the Kalman filter. *IEEE Trans. Aerospace and Elec. Sys.,*Vol. AES-23, No. 2, 1987.

9. Chen, C.Y. and Abraham, J.A., Fault-tolerant systems for the computation of eigen-values and singular values. *SPIE Vol. 696 Adv. Alg. and Arch. for Signal Proc.*, pp. 228–237, 1986.

10. Mitchell, W.F., A fault tolerant Kalman filter systolic array. General Electric internal report, 1990, *SIAM J. Mat. Anal. and App.*, submitted.

11. Mitchell, William F., *An N(N +1) Processor 4N +6M Step Kalman Filter Systolic Array*, Advanced Technology Laboratories, General Electric Company, Moorestown, NJ; Technical Report CMAT-90 TR-007.

9.6   APPLICATIONS OF A PARALLEL PRESSURE-CORRECTION ALGORITHM TO THREE-DIMENSIONAL TURBOMACHINERY FLOWS

A parallel algorithm for the solution of three-dimensional compressible flows in turbomachinery has been developed and demonstrated on scalable distributed memory multicomputers [1]. The algorithm solves the compressible form of the Euler or Navier-Stokes equations via a compressible pressure correction formulation. To achieve high accuracy for highly turning blade rows, the computational grid is constructed without requiring strict periodicity of the grid points along the periodic boundaries between the blade passages. The impact of this feature on code parallelization and computational efficiency is described. The algorithm has been demonstrated on up to 128 processors of an Intel iPSC/860, and up to 256 processors of the Intel Touchstone Delta prototype (Paragon). Performance 2.5 times faster than a single Cray Y-MP processor has been achieved for an inviscid turbomachinery calculation on 154,000 grid points with 256 processors of the Delta.

The research objective of this effort is to develop a parallel turbomachinery code capable of running on massively parallel computers at greater performance levels than are obtainable on conventional vector supercomputers. This capability will allow larger, more accurate simulations of viscous transonic turbomachinery flows than is now possible.

Traditionally, time-marching methods such as Jameson's explicit Runge-Kutta scheme [2] or the Beam-Warming implicit scheme [3] have been used to solve compressible turbomachinery flows. In recent years, pressure correction algorithms, originally developed to handle incompressible flows, have been successfully extended to handle compressible flows. A series of papers [4–6], described the development of a two-dimensional parallel pressure correction algorithm applicable to both incompressible and compressible flow.

The compressible pressure correction algorithm described in reference [6] has been extended to three dimensions, and calculations for actual turbomachinery blade rows used in modern gas turbine engines have been performed on a large scalable distributed memory multicomputer. Complexities in the algorithm required for the proper treatment of highly turning blade rows have been success-fully addressed in the parallel implementation. High parallel efficiency and performance in excess of modern single processor supercomputers have been achieved.

The compressible pressure correction equation developed here solves the three-dimensional Euler equations for inviscid flow, or the fully elliptic form of the Navier-Stokes equations for viscous flows. The equations for conservation of x-, y-, and z-momentum, mass, and enthalpy are solved, along with the ideal gas equation of state. For turbulent flows, the standard k-e turbulence model is used, along with the wall function treatment for the near-wall regions.

The conventional pressure correction formulation [7], originally developed for incompressible flows, derives an equation for the pressure correction by ma-nipulation of the discrete forms of the momentum and continuity equations. The momentum equations are first solved using a guessed pressure field. The pressure correction equation serves both to correct the velocities to enforce continuity and to provide an updated pressure field. During this process, the density field is taken as fixed.

In the compressible pressure correction formulation, both the velocity and density fields are simultaneously updated to enforce continuity. The resulting al-gorithm has the attractive property of being able to address flows at all Mach numbers, making it very widely applicable.

Upwinding of the densities provides the mechanism for shock capturing in transonic and supersonic flows. The conservative second-order accurate QUICK (quadratic upstream differencing) scheme [8] is used to compute the combined convection-diffusion fluxes. Although the first-order accurate hybrid differencing scheme is still widely used for incompressible flows, it leads to excessive smear-ing when shocks are present and excessive total pressure errors. Shocks are found to be captured within three to four grid cells when QUICK is used, and total pres-sure conservation in inviscid flows is found to be significantly better.

Highly turning turbine blade rows pose difficulties to the conventional H-grids commonly used for turbomachinery calculations. The requirement of strict grid periodicity, at the periodic boundaries causes severe grid shearing, and loss of orthogonality of the grid at the blade surface. This grid shearing can lead to significant numerical error, and loss of stability of the solution algorithm. Lack of grid orthogonality in the near-wall region also reduces the accuracy of the wall functions used to model the turbulent boundary layer in viscous cases.

A remedy for this problem is to relax the requirement of strict grid periodic-ity, and allow for a mismatch of the grid points across the periodic boundary. The resulting grid is much less sheared, and grid orthogonality near the wall is much improved. The penalty in this approach is the need to interpolate values commu-nicated across the periodic boundary, and the interpolation errors that conse-quently arise. These errors can be reduced by taking the periodic boundary to be at the mid-passage line so that the interpolations are performed away from the leading and trailing edges of the blade, where the flow gradients are typically highest. This requires the blade to be meshed as part of the computational domain. Here the blade is treated as a slit in the transformed domain, resulting in a "triple grid line" both upstream and downstream of the trailing edge of the blade, where three grid lines collapse together. The presence of triple grid lines requires special treatment of the zero-volume cells that result, as described in [9].

The natural means of parallelizing the basic algorithm is by domain decomposition. A full three-dimensional decomposition is done by dividing the solution domain into overlapping cubical regions (in the transformed space), such that an equal number of internal control volumes are assigned to each processor. This ensures a reasonable load balance since the number of floating point operations per control volume is roughly equal. The desired number of processors in each of the three grid directions can be set by the user at run time. An overlap of one cell in each direction at each face is used to minimize the redundant storage of quantities in the overlapping regions. Six temporary arrays are used to provide storage for the additional planes of values required by the QUICK scheme, rather than by using a two cell overlap. Additional communication is required across the periodic boundaries and across any triple grid lines present, to update the required values. The communication across the periodic boundaries requires special considerations, as described later, but the communication across the triple grid line is straightforward.

The algorithm developed here uses a standard staggered grid formulation for the velocity components to prevent problems with spurious oscillations in the computed pressure field. In the staggered grid scheme, the velocity components are solved on the faces of the control volumes, whereas the other scalar variables such as pressure correction and temperature are computed at the control volume centers. In a given grid direction, there is always one less staggered velocity component to be solved for than there are corresponding scalar values. Since the solution of the pressure correction equation takes up the majority of the computation time, the best domain decomposition leads to an equal number of scalar grid points (or equivalently, control volumes) on each processor. This ensures a good load balance for the solution of this (and the other scalar) equations. Unfortunately, such a decomposition always leaves one processor with one less velocity value in each line to solve than the others, which results in a computational load imbalance during the solution of the momentum equations. This load imbalance is a major contributor to parallel inefficiency for this algorithm, a fact that suggests that unstaggered grid formulations may ultimately prove superior for parallel machines.

The possible presence of mismatched grid points at the periodic boundary has important implications on the parallel implementation of the algorithm. The resulting communication across the periodic boundary is not necessarily one-to-one between processors on opposite sides of the domain. Rather, it depends on the particular nature of the grid. The situation is analogous to the domain decomposition of an unstructured grid, in which each processor can have different numbers of neighbors, and different amounts of data that need to be communicated with different processors.

The approach taken here was to stay with a regular structured three-dimensional domain decomposition and suffer the inefficiency that results in the transfer of values across the periodic boundary. This approach was chosen because the exchange of values takes very little time relative to the bulk of the calculation for the serial code. At the beginning of the calculation, each processor builds lists of processors that periodic boundary values will be sent to and received from. First, the entire plane of global coordinates on the far side of periodic boundary is temporarily stored on each processor, and then triangulated. For each point on its side of the periodic boundary, the processor determines the corresponding triangle on the far side, the processor the triangle belongs to, the three vertex locations, and the corresponding area coordinates. These values are assembled into lists, which are subsequently used to pass and interpolate values across the periodic boundary.

Completing the parallelization of these routines proved to be a difficult job because of the extensive bookkeeping involved, and consumed 3 weeks of effort. Happily, the inefficiency caused by the slight communication imbalance across the mismatched periodic boundary had no significant impact on the parallel efficiency of the overall application for reasonably large grids.

A series of calculations for inviscid and viscous flow in turbomachinery blade rows has been completed. The problems studied included a test case comprised of a cascade of biconvex airfoils, a two-dimensional transonic turbine test rig, a highly turning subsonic turbine nozzle from an industrial steam turbine, and a highly turning transonic turbine nozzle from an industrial gas turbine. The discussion will focus on the results from the test rig and show only timings for the other cases.

All parallel computations shown were performed in 32-bit arithmetic on a 16-node Intel iPSC/860, a 128-node iPSC/860 at the NASA-Ames Research Center, and the 512-node Intel Touchstone Delta prototype at the California Institute of Technology. Experience with this algorithm indicates that 64-bit arithmetic is not required. However, Cray timings do reflect 64-bit arithmetic since that is the default on that machine.

The parallel algorithm was first tested for the inviscid flow through a cascade of biconvex airfoils, computed on grid with 30,000 grid points. These calculations were done using the Green Hills compiler on the NASA-Ames iPSC/860 machine. With 64 processors, the performance of the iPSC/860 was 40% faster than the vectorized production code running on a single processor of a Cray Y-MP. With the full scalar optimization of the Green Hills FORTRAN compiler, performance 10% better than the Cray was achieved with only 32 processors.

Table 9-7 shows measured and calculated performance for a turbulent viscous calculation on a highly running transonic turbine nozzle, computed on a grid with 37,000 grid points. These calculations were done using the pipelining pgf77 compiler. With 16 processors, performance equivalent to 79% of a single Cray Y-MP processor was achieved, along with an estimated parallel efficiency of 80.5%.

Table 9-8 shows measured performance for the calculation of an inviscid transonic turbine cascade on a grid with $130 \times 18 \times 66$ points. Since access to the full complement of processors on either the Ames machine or the Delta machine is limited, the grid was generated to be as uniform in spacing as possible, since this helps ensure reliable convergence of the algorithm, at the expense of grid resolution around the leading and trailing edges of the blade. Although this compromises the accuracy of the computed results somewhat, the timings, which are of principal interest, are valid. When parallel machines of this class become more widely available, there will be plenty of opportunity to repeat the calculation on grids with more nonuniform spacing which more accurately resolve the solution.

Comparison of the results was made to timings on a Cray Y-MP running a highly vectorized production version of the original serial algorithm. Performance 2.4 times faster than a single processor of a Cray Y-MP was achieved with 128 processors of the iPSC/860, and performance 2.5 times faster was achieved with 256 processors of the Delta machine.

Since this test case is too large to run on a single processor, the parallel efficiency must be estimated. Here, this was done using the method described in [5]. The parallel efficiency drops below 50% around the level of 64 processors, suggesting that the given problem is too small for a much greater number of processors. The loss of parallel efficiency is due primarily to load imbalances caused by the use of the staggered grid (as discussed earlier), and from the application of different boundary conditions on different processors, as well as the cost of communications between processors during the solution step.

TABLE 9-7

PERFORMANCE OF PARALLEL ALGORITHM ON VISCOUS TRANSONIC TURBINE CASCADE*			
# Processors	Sec/Iter/ Point/Equation	Performance/Cray	Parallel Efficiency (%)
Cray Y-MP (1)	1.1×10^{-5}	1.00	—
iPSC/860 (8)	2.5×10^{-5}	0.44	89.9
iPSC/860 (16)	1.4×10^{-5}	0.79	80.5

*37,000 grid points.

TABLE 9-8

PERFORMANCE OF PARALLEL ALGORITHM ON INVISCID TRANSONIC TURBINE CASCADE*			
# Processors	Sec/Iter/ Point/Equation	Performance/Cray	Parallel Efficiency (%)
Cray Y-MP (1)	1.1×10^{-5}	1.00	—
iPSC/860			
$4 \times 2 \times 2$ (16)	1.51×10^{-5}	0.73	80.9
$4 \times 2 \times 4$ (32)	8.85×10^{-6}	1.24	67.1
$4 \times 4 \times 4$ (64)	6.29×10^{-6}	1.75	50.1
$8 \times 4 \times 4$ (128)	4.59×10^{-6}	2.40	33.3
Touchstone Delta			
$4 \times 2 \times 2$ (16)	1.36×10^{-5}	0.81	74.8
$4 \times 2 \times 4$ (32)	8.09×10^{-5}	1.36	58.9
$4 \times 4 \times 4$ (64)	5.84×10^{-6}	1.89	41.4
$8 \times 4 \times 4$ (128)	4.47×10^{-6}	2.46	25.9
$8 \times 4 \times 8$ (256)	4.30×10^{-6}	2.56	14.9

*154,000 grid points.

Parallel efficiency for this problem appears slightly poorer on the Delta machine than on the iPSC/860, for the same number of processors. Although at first this result seems surprising, since the Delta is supposed to have much better communication bandwidth than the iPSC/860, on closer inspection it can be explained. First, the i860 processors on the Delta machine are a later revision than those on the Ames machine, and are reportedly 10 to 15% faster. Faster processors lead to a higher communication/computation ratio, if the speed of the underlying communication network remains the same. Most of the local messages passed by this algorithm are of only moderate length (typically, 1000 bytes), so latency effects are still important, and increased bandwidth is not significant.

The global communications required by the code were also found to be somewhat slower on the Delta machine than on the iPSC/860. The global communications routines used in the code were developed prior to Intel's offering of library routines such as GSSUM, and are based on the hypercube communications algorithms described in [10]. These algorithms help assure contention-free communication on a hypercube since the messages travel between nearest neighbors along one dimension of the hypercube at a time as the global operation proceeds. The time required to complete the global operation scales as log P, where P is the number of processors in the hypercube. On the iPSC/860, the performance of the handwritten routines is similar to that of the GSSUM library routine, with a slight edge to GSSUM as the number of processors becomes large.

On the mesh topology of the Delta, however, the situation is very different. The hypercube-based algorithm does not lead to nearest neighbor communication, and the logarithmic scaling does not hold. The effects of contention become very apparent beyond the level of 64 processors. The Intel library routine GSSUM, which implements an appropriate mesh-based algorithm on the Delta, shows comparable performance on the Delta to GSSUM on the IPSC/860, and clearly outperforms the hand-written routine.

There are some lessons to be learned from this experience. Details of the underlying topology do influence the communication performance, particularly for global communications where all of the processors participate simultaneously, loading up the network with messages. The standard practice in writing parallel codes for earlier hypercube machines was to extensively optimize the code for the hypercube architecture using such devices as gray codes, hypercube-based shuffle and global summation routines, and so on. These optimizations, which were specific to the details of the underlying topology of the machine, can be detrimental to performance when the code is ported to a machine with a different underlying topology.

A better practice is to define some standard global communication operations, such as global summation, with standard calls, and let the computer vendors implement these operations on their machines in whatever manner is most efficient for their particular machine architecture. The GSSUM routine provided by Intel has the same syntax on both the iPSC/860 and Delta, so the coding is truly portable, and the performance on both machines is comparable. This is in significant contrast to the optimized hypercube-based routine, which although portable, showed a significant performance loss on the mesh topology of the Delta.

ACKNOWLEDGMENTS

Thanks are offered ICOMP at NASA-Lewis and NAS at NASA-Ames for access to the iPSC/860 at NASA-Ames. Thanks are also due to NASA-Lewis for providing access to the Intel Touchstone Delta System operated by Caltech on behalf of the Concurrent Supercomputing Consortium.

REFERENCES

1. Braaten, Mark E., *Applications of a Parallel Pressure-Correction Algorithm to 3D Turbomachinery Flows*, (GE Research and Development Center, Schenectady, NY 12031), Parallel Computational Fluid Dynamics 92,

Proceedings of the Conference on Parallel Computational Fluid Dynamics, Hauser, J., ed., Princeton, May 1992.

2. Jameson, A., Schmidt, W., and Turkel, E., Numerical Solution of the Euler Equations by Finite Volume Methods Using Range-Kutta Stepping Schemes, AIAA paper AIAA-81-1259, 1981.

3. Beam, R.W. and Warming, R.F., An Implicit Finite Difference Algorithm for Hyperbolic System in Conservation Form. *J. Comp. Phys.*, Vol. 23, pp. 87–110, 1976.

4. Bratten, M.E., Solution of Viscous Fluid Flows on a Distributed Memory Concurrent Computer. *Int. J. Num. Meth. Fluids,* Vol. 10, pp. 889–905, 1990.

5. Bratten, M.E., Development of a Parallel Computational Fluid Dynamics Algorithm on a Hypercube Computer. *Int. J. Num. Meth. Fluids,* Vol. 12, pp. 947–963, 1991.

6. Bratten, M.E., Parallel Computation of the Compressible Navier-Stokes Equations with a Pressure-Correction Algorithm, in Walker, D.W., and Stout, Q.F., eds., *Proceedings of the Fifth Distributed Memory Computing Conference,* April 8–12, 1990, Charleston, SC, pp. 463–469, 1990.

7. Patankar, S.V., Numerical Heat Transfer and Fluid Flow, Hemisphere, Washington, DC, 1980.

8. Leonard, B.P., A Stable and Accurate Convective Modeling Procedure Based on Quadratic Upstream Interpolation. *Comput. Meths. Appl. Mech. Eng.,* Vol. 19, pp. 59–98, 1979.

9. Cedar, R.D. and Holmes, D.G., The Calculation of Three-dimensional Flow through a Transonic Fan Including the Effects of Blade Surface Boundary Layers, Partspan Shroud, Engine Splitter and Adjacent Blade Rows, AMSE paper AMSE 89-GT-325, 1989.

10. Saad, Y. and Schultz, M.H., Data Communication in Hypercubes. Report YALEU/DCS/RR-248, Yale University, 1985.

Appendix

This section contains a series of serial programs. For each program there is also a parallel version using PVM and a MAKEFILE to build the executable. Both serial and parallel versions are in FORTRAN.

SIMPLE ARRAY ASSIGNMENT [1]

Description: This is the "hello world" of parallel programming. It is a simple array assignment used to demonstrate the distribution of data among multiple tasks and the communications required to accomplish it. In the parallel version, the master task distributes an equal portion of the array to each worker task. Each worker task receives its portion of the array and performs a simple value assignment to each of its elements. Each worker then sends its portion of the array back to the master. As the master receives back each portion of the array selected elements are displayed.

SERIAL FORTRAN VERSION
OF ARRAY ASSIGNMENT PROGRAM

```
C***************************************************************************
C Serial Example - Array Assignment - Fortran Version
C FILE: ser_array.f
C OTHER FILES: make.array.f
C DESCRIPTION:
C   This is a simple array initialization and assignment.  In the parallel
C   version, the master task initiates numtasks-1 number of
C   worker tasks.  It then distributes an equal portion of an array to each
C   worker task.  Each worker task receives its portion of the array, and
```

```
C    performs a simple value assignment to each of its elements.  The value
C    assigned to each element is simply that element's index in the array+1.
C    Each worker task then sends its portion of the array back to the master
C    task.  As the master receives back each portion of the array, selected
C    elements are displayed.
C AUTHOR: Blaise Barney
C LAST REVISED:  4/18/94 Blaise Barney
C************************************************************************

      program array

      integer   ARRAYSIZE
      parameter (ARRAYSIZE = 600000)

      integer   index, i
      real*4    data(ARRAYSIZE)


      print *, '********** Starting SERIAL Example ************'

C    Initialize the array
      do 20 i=1, ARRAYSIZE
        data(i) = 0.0
 20   continue

      do 50 i=1, ARRAYSIZE
        data(i) = i + 1
 50   continue

C    Print some results

      print *, '- - - - - - - - - - - - - - - - - - - - - -'
      print *, ' Sample Results '
      print *, 'data[',1, ']=', data(1)
      print *, 'data[',100, ']=', data(100)
      print *, 'data[',1000, ']=', data(1000)
      print *, ' All Done!'

      end

#########################################################################
# Serial  Array Example Makefile
# FILE: make.ser_array.f
# DESCRIPTION:  See ser_array.f
# USE: make -f make.ser_array.f
# LAST REVISED: 4/18/94 Blaise Barney
#########################################################################
```

```
F77          =        xlf

array:          ser_array.f
        ${F77} ser_array.f  -o array
```

PARALLEL EXAMPLE OF ARRAY ASSIGNMENT PROGRAM IN FORTRAN USING PVM FOR INTERPROCESSOR COMMUNICATION: MASTER PROGRAM

```
C**********************************************************************
C PVM Example - Array Assignment Fortran Version
C Master Program
C FILE: pvm_array.master.f
C OTHER FILES: pvm_array.worker.f  make.pvm_array.f
C DESCRIPTION: Fortran Language version of PVM example 1 master task
C    In this simple example, the master task initiates a specified number of
C    instances of the worker task.  It then distributes an equal portion of
C    an array to each instance of the worker task.  Each instance of the
C    worker task receives its portion of the array, and performs a simple
C    value assignment to each of its elements. The value assigned to each
C    element is simply that element's index in the array+1.  Each worker task
C    then sends its portion of the array back to the master task.  As the
C    master receives back each portion of the array, selected elements are
C    displayed. Note that the order in which the worker tasks finish is
C    non-deterministic so that different portions of the array may display
C    "out of order".
C PVM VERSION: 3.x
C AUTHOR: Blaise Barney
C LAST REVISED: 4/18/94 Blaise Barney
C**********************************************************************

        program array_master

C PVM Version 3.0 include file
        include 'fpvm3.h'

        integer   NTASKS, ARRAYSIZE
        parameter (NTASKS = 6)
        parameter (ARRAYSIZE = 600000)
        parameter (FROMMASTER_MSG      = 1)
        parameter (FROMWORKER_MSG      = 2)
```

```
       integer     tids(NTASKS), rc, i, index, tid,
    &               bufid, bytes, msgtype, chunksize
       real*4      data(ARRAYSIZE), result(ARRAYSIZE)

C*********************** enroll this task in PVM ***********************
C  pvmfmytid will enroll this process in your PVM virtual machine.  A unique
C  task id will be assigned if call is successful.  The pvmds keep track of
C  processes and communications via task ids.  Return codes less than zero
C  indicate an error in the enroll process and will terminate this program.
C**********************************************************************
       print *, '********** Starting PVM Example ************'
       call pvmfmytid(rc)
       if (rc .lt. 0) then
         print *, 'MASTER: Unable to enroll this task.'
         print *, '   Return code= ', rc, '. Quitting.'
         stop
       else
         print *, 'MASTER: Enrolled as task id = ', rc
       endif
C************************* spawn worker tasks ***************************
C  The master task now spawns the worker tasks by calling pvmfspawn. The unique
C  task ids for workers are stored in the tids array.  The return code tells
C  the number of tasks successfully spawned.  In this example, it must equal
C  NTASKS, otherwise the program terminates.
C**********************************************************************
       print *, 'MASTER: Spawning worker tasks. . .'
       call pvmfspawn("array.worker", PVMDEFAULT, " ", NTASKS,
    &                tids, rc)
       if (rc .eq. NTASKS) then
         print *, 'MASTER: Successfully spawned ', rc, ' worker tasks.'
       else
         print *, 'MASTER: Not able to spawn requested number of tasks!'
         print *, 'MASTER: Tasks actually spawned: ', rc, ',  Quitting,'
         stop
       endif


C*********************** initializations ****************************
C  Define the partition size as chunksize and then initialize the array to 0
C**********************************************************************
       chunksize = (ARRAYSIZE / NTASKS)
       do 20 i=1, ARRAYSIZE
         data(i) =  0.0
 20    continue


C**************** send array chunks to each worker task **************
C  Data passing from master task to each worker task begins by initializing the
C  send buffer with a call to pvmfinitsend.  Its argument tells PVM that XDR
```

```
C  data conversion should be performed only if the virtual machine is hetero
C  genous.  Data is packed sequentially into the buffer with the calls to
C  pvmfpack.  The arguments to pvmfpack specify the data type, data
C  address, number of values and stride.  Each worker is sent the following:
C     index = the starting index for this worker's partition of the array;
C               one value is sent with a stride on one.
C     chunksize = the partition size of the array, one value is sent with a
C                 stride of one.
C     data() = the actual data for this worker's array partition; chunksize
C                 number of values are sent with a stride of one.
C  Finally, the data is sent to each worker task by calling pvmfsend.  Its
C  arguments specify each task id that is to receive the data and which
C  message type should be set.  Index is incremented by chunksize for the
C  next workers partition of the array.  Note that PVM routine calls provide
C  return codes even if they are not always used in this program.
C*****************************************************************************
        print *, 'MASTER: Sending data to worker tasks...'
        index = 1
        msgtype = FROMMASTER_MSG
        dO 30 i=1, NTASKS
           call pvmfinitsend(PVMDEFAULT, rc)
           call pvmfpack(INTEGER4, index, 1, 1, rc)
           call pvmfpack(INTEGER4, chunksize, 1, 1, rc)
           call pvmfpack(REAL4, data(index), chunksize, 1, rc)
           call pvmfsend(tids(i), msgtype, rc)
           index = index + chunksize
 30     continue


C*************** wait for results from all worker tasks ********************
C  The master task now waits in a loop to receive each worker's partition of the
C  array.  pvmfrecv blocks until it receives a message of type "msgtype"
C  from task id = -1, which means, "any task id". pvmfrecv returns the PVM
C  message buffer id of the awaited message as bufid. Bufid is then used to
C  find out additional information about the message by calling pvmfbufinfo.
C  The data in the message buffer is unpacked sequentially with calls to
C  pvmfunpack.  The arguments to pvmunpack specify the data type,
C  address of the data, number of values and stride.  The master knows which
C  part of the result array it is receiving by the value of index.  As above
C  the size of the array partition is determined by chunksize.  Note that it
C  is the programmer's responsibility to insure that message types and data
C  sequences match between data sends/receives. Finally, the master task
C  prints a sample of the returned result partition and also tells which
C  task id it came from by the value of tid, as returned from the call to
C  pvmfbufinfo.
C*****************************************************************************
        print *, 'MASTER: Waiting for results from worker tasks...'
        msgtype = FROMWORKER_MSG
```

```
      do 40 i=1, NTASKS
        call pvmfrecv(-1, msgtype, bufid)
        call pvmfbufinfo(bufid, bytes, msgtype, tid, rc)
        call pvmfunpack(INTEGER4, index, 1, 1, rc)
        call pvmfunpack(REAL4, result(index), chunksize, 1, rc)
        print *, '- - - - - - - - - - - - - - - - - - - - - - - - -'
        print *, 'MASTER: Sample results from worker task id = ', tid
        print *, '   result[', index, ']=', result(index)
        print *, '   result[', index+100, ']=', result(index+100)
        print *, '   result[', index+1000, ']=', result(index+1000)
        print *, ' '
 40   continue


C************************* exit from PVM *************************************
C  Leave PVM before program finishes.  This is "good programming practice"
C  since it allows all of the pvmds in your virtual machine to keep track of
C  which tasks are active and which are not.  A return code less than zero
C  from pvmfexit indicates an error which is ignored by this program.
C****************************************************************************
      print *, 'MASTER: All Done!'
      call pvmfexit(rc)
      end
```

PARALLEL EXAMPLE OF ARRAY ASSIGNMENT PROGRAM IN FORTRAN USING PVM FOR INTERPROCESSOR COMMUNICATION: WORKER PROGRAM

```
C****************************************************************************
C PVM Example - Array Assignment Fortran Version
C Worker Program
C FILE: pvm_array.worker.f
C DESCRIPTION: See pvm_array.master.f
C PVM VERSION: 3.x
C LAST REVISED: 4/18/94 Blaise Barney
C****************************************************************************

      program array_worker

C PVM version 3.0 include file
      include 'fpvm3.h'
```

```fortran
      integer ARRAYSIZE
      parameter(ARRAYSIZE      = 600000)
      parameter(FROMMASTER_MSG    = 1)
      parameter(FROMWORKER_MSG    = 2)


      integer     i, masterid, rc, index, msgtype, chunksize
      real*4      result(ARRAYSIZE)


C********************** enroll this task in PVM ***************************
C  pvmfmytid will enroll this process in your PVM virtual machine.  A unique
C  task id will be assigned if call is successful.  The pvmds keep track of
C  processes and communications via task ids.  Return codes less than zero
C  indicate an error in the enroll process and will terminate this program.
C  Note that worker programs, though spawned from the master, must still
C  enroll.
C*********************************************************************************
      call pvmfmytid(rc)
      if (rc .lt. 0) then
        print *, 'WORKER: Unable to enroll this task.'
      print *, '    Return code = ', rc, '.  Quitting.'
        stop
      else
        print *, 'WORKER: Enrolled as task id = ', rc
      endif


C***************** receive data from master task ***************************
C  The worker task blocks with pvmfrecv until it receives a message from its
C  parent - the master task - of the correct message type.  It knows the task
C  id of its parent from the call to pvmfparent.  When the correct message
C  is received, the data is unpacked by calls to pvmfunpack(INTEGER4/REAL4)
C  in the identical sequential order that it was originally sent from the
C  master.  Note that it is the programmer's responsibility to insure that
C  message types and data sequences match between data sends/receives.
C*********************************************************************************
      msgtype = FROMMASTER_MSG
      call pvmfparent(masterid)
      call pvmfrecv(masterid, msgtype, rc)
      call pvmfunpack(INTEGER4, index, 1, 1, rc)
      call pvmfunpack(INTEGER4, chunksize, 1, 1, rc)
      call pvmfunpack(REAL4, result(index), chunksize, 1, rc)


C*********** modify the array before sending it back to master ************
C  Real complex operation going on here - the worker just adds one to the
C  index value of the array location.
C*********************************************************************************
      do 10 i=index, index + chunksize
        result(i) = i + 1
```

```
 10    continue

C******************* send results back to master task **********************
C  Data passing from this worker task to master task begins by initializing
C  the send buffer with a call to pvmfinitsend.  Its argument tells PVM
C  that XDR data conversion should be performed only if the virtual machine
C  is heterogenous.  Data is packed sequentially into the buffer with the
C  calls to pvmfpack.  The arguments to pvmfpack specify the data type, data
C  address, number of values and stride.  Each worker sends the following:
C     index = the starting index for this worker's partition of the array;
C              one value is sent with a stride on one.
C     result() = the actual data for this worker's array partition; chunksize
C              number of values are sent with a stride of one.
C  Finally, the data is sent by the worker task by calling pvmfsend.  Its
C  arguments specify the task id of the task which should receive the data -
C  in this case the master task id which it obtained previously - and which
C  message type should be set.  Note that PVM routine calls provide return
C  codes even if they are not always used in this program.
C**************************************************************************
        msgtype = FROMWORKER_MSG
        call pvmfinitsend(PVMDEFAULT, rc)
        call pvmfpack(INTEGER4, index, 1, 1, rc)
        call pvmfpack(REAL4, result(index), chunksize, 1, rc)
        call pvmfsend(masterid, msgtype, rc)


C********************** exit from PVM **************************************
C  Leave PVM before program finishes.  This is "good programming practice"
C  since it allows all of the pvmds in your virtual machine to keep track of
C  which tasks are active and which are not.  A return code less than zero
C  from pvmfexit indicates an error which is ignored by this program.
C**************************************************************************
        call pvmfexit(rc)
        end
# # # # # # # # # # # # # # # # # # # # # # # # # # # # # # # # # # # # # #
# PVM Array Makefile
# FILE: make.pvm_array.f
# DESCRIPTION:  Makefile for pvm array example.  Fortran Language
# AUTHOR:  Blaise Barney
# PVM VERSION: 3.x
# USE: make -f make.pvm_array.f
# LAST REVISED: 4/18/94 Blaise Barney
# # # # # # # # # # # # # # # # # # # # # # # # # # # # # # # # # # # # # #
F77       = xlf
MASTER    = array.master
MASTERSRC = pvm_array.master.f
WORKER    = array.worker
WORKERSRC = pvm_array.worker.f
```

```
PVMDIR    = ${WORKSHOP}/pvm3
INCLUDE   = -I${PVMDIR}/include
LIBS      = -L${PVMDIR}/lib/RS6K -lfpvm3 -lpvm3


doit:     ${MASTER} ${WORKER}


${MASTER}: ${MASTERSRC}
     $(F77) ${MASTERSRC} ${INCLUDE} ${LIBS} -o ${MASTER}


${WORKER}: ${WORKERSRC}
     ${F77} ${WORKERSRC} ${INCLUDE} ${LIBS} -o ${WORKER}
```

MATRIX MULTIPLY [2]

Description: This example is a simple matrix multiply program. In the parallel version, the data are distributed among the worker tasks who perform the actual multiplication and send back their respective results to the master task.

SERIAL MATRIX MULTIPLY PROGRAM
IN FORTRAN

```
C******************************************************************************
C SERIAL MATRIX MULTIPLY - Fortran Version
C FILE: ser_mm.f
C OTHER FILES: make.ser_mm.f
C DESCRIPTION: This is the serial version of the matrix multiply example.
C    To make this a parallel processing program this program would be divided
C    into two parts - the master and the worker section.  The master task
C    would distributes a matrix multiply operation to numtasks-1  worker
C    tasks. NOTE1:  C and Fortran versions of this code differ because of
C    the way arrays are stored/passed.  C arrays are row-major order but
C    Fortran arrays are column-major order.
C AUTHOR: Ros Leibensperger / Blaise Barney
C LAST REVISED:   03/21/94 Made serial version - R. Arnowitz
C******************************************************************************

      program mm

      parameter (NRA = 62)
      parameter (NCA = 15)
      parameter (NCB = 7)
```

```fortran
      integer          cols,avecol,i,j,k
      real*8 a(NRA,NCA), b(NCA,NCB), c(NRA,NCB)


C     Initialize A and B
      do 30 i=1, NRA
        do 30 j=1, NCA
        a(i,j) = (i-1)+(j-1)
 30   continue
      do 40 i=1, NCA
        do 40 j=1, NCB
      b(i,j) = (i-1)*(j-1)
 40   continue


C     Do matrix multiply
       do 50 k=1, NCB
         do 50 i=1, NRA
         c(i,k) = 0.0
           do 50 j=1, NCA
           c(i,k) = c(i,k) + a(i,j) * b(j,k)
 50        continue


C     Print results
      print*, '"Here is the result matrix: '
      do 90 i=1, NRA
        do 80 j = 1, NCB
        write(*,70)c(i,j)
 70     format(2x,f8.2,$)
 80     continue
      print *, ' '
 90   continue
end
```

```makefile
# # # # # # # # # # # # # # # # # # # # # # # # # # # # # # # # # # # # # # # #
# Serial mm Example Makefile
# FILE: make.ser_mm.f
# DESCRIPTION:  See ser_mm.f
# USE: make -f make.ser_mm.f
# LAST REVISED: 4/18/94 Blaise Barney
# # # # # # # # # # # # # # # # # # # # # # # # # # # # # # # # # # # # # # # #
F77       =     xlf

mm:     ser_mm.f
      ${F77} ser_mm.f  -o mm
```

PARALLEL MATRIX MULTIPLY PROGRAM IN FORTRAN USING PVM FOR INTERPROCESSOR COMMUNICATION: MASTER PROGRAM

```
C*************************************************************************
C PVM Matrix Multiply - Fortran Version
C Master Program
C FILE: pvm_mm.master.f
C OTHER FILES: pvm_mm.worker.f, make.pvm_mm.f
C DESCRIPTION:  PVM matrix multiply example code master task.  Fortran version.
C   In this example code, the master program acts as the parent and spawns
C   NPROC worker tasks.  The first worker task is spawned on a specific
C   machine. The master program performs the matrix multiply by sending all of
C   matrix A to every worker task and then partitioning columns of matrix B
C   among the workers. The worker tasks perform the actual multiplications
C   and send back to the master task their respective results.
C   NOTE1:  C and Fortran versions of this code differ because of the way
C   arrays are stored/passed.  C arrays are row-major order but Fortran
C   arrays are column-major order.
C PVM VERSION: 3.x
C AUTHOR: Blaise Barney - adapted from C version
C LAST REVISED: 4/18/94 Blaise Barney
C*************************************************************************
C Explanation of constants and variables used in this program:
C   NPROC            = number of PVM worker tasks to spawn
C   NRA              = number of rows in matrix A
C   NCA              = number of columns in matrix A
C   NCB              = number of columns in matrix B
C   mtid             = PVM task id of master task
C   wtids                = array of PVM task ids for worker tasks
C   mtype                = PVM message type
C   cols             = columns of matrix B sent to each worker
C   avecol, extra    = used to determine columns sent to each worker
C   offset           = starting position within the matrix
C   rcode, i, j          = misc.
C   a                = matrix A to be multiplied
C   b                    = matrix B to be multiplied
C   c                = result matrix C
C   thishost         = name of selected master
C - - - - - - - - - - - - - - - - - - - - - - - - - - - - - - - - - - - -
```

```
      program mm_master
C     PVM Version 3.0 include file
      include 'fpvm3.h'

      parameter (NPROC = 4)
      parameter (NRA = 62)
      parameter (NCA = 15)
      parameter (NCB = 7)

      integer     mtid, wtids(NPROC), mtype, cols, avecol, extra, offset,
     &     rcode, i, j
      real*8   a(NRA,NCA), b(NCA,NCB), c(NRA,NCB)
      character*35 thishost

C  Enroll this task in PVM
      call pvmfmytid(mtid)

C  The master task now spawns worker tasks by calling pvm_spawn.  The unique
C  worker task ids are stored in the wtids array.  The first worker task is
C  spawned on a specific machine.  The return code tells the number of tasks
C  successfully spawned, and in this example, is not checked for errors.
      Do 20 i=1, NPROC
      if (i .eq. 1) then
        write(*,9)
9       format('Enter selected hostname - must match PVM config: ',$)
        read (*, 10) thishost
10      format (a35)
        call pvmfspawn("mm.worker",PVMHOST,thishost,1,wtids(1),rcode)
      else
        call pvmfspawn("mm.worker", PVMDEFAULT, " ", 1, wtids(i), rcode)
      endif
20    continue


C  Initialize A and B
      do 30 i=1, NRA
        do 30 j=1, NCA
        a(i,j) = (i-1)+(j-1)
30    continue

      do 40 i=1, NCA
        do 40 j=1, NCB
      b(i,j) = (i-1)*(j-1)
40    continue
```

```
      avecol = NCB/NPROC
      extra = mod(NCB, NPROC)
      offset = 1
      mtype = 1

C  Send data to the worker tasks
C  First find #columns from B to send to each worker task
      Do 50 i=1, NPROC
      if (i .le. extra) then
        cols = avecol + 1
      else
        cols = avecol
      endif

C  Next call initializes send buffer and specifies to do XDR data format
C  conversion only in heterogenous environment
      call pvmfinitsend(PVMDEFAULT, rcode)

C  Next four calls pack values into the send buffer - rcode not checked
C      offset   = starting position in matrix
C      cols     = number of columns of B to send
C      a                  = send all of A
C      b                  = send some columns from B beginning at offset
      call pvmfpack(INTEGER4, offset, 1, 1, rcode)
      call pvmfpack(INTEGER4, cols, 1, 1, rcode)
      call pvmfpack(REAL8, a, NRA*NCA, 1, rcode)
      call pvmfpack(REAL8, b(1,offset), cols*NCA, 1, rcode)

C  Send contents of send buffer to worker task
      call pvmfsend(wtids(i), mtype, rcode)

      offset = offset + cols
  50  continue

C  Wait for results from all worker tasks.  After setting message type,
C  loop for NPROCs.  Receive following data from each worker:
C      offset          = starting position in matrix
C      cols            = number of columns to receive
C      c(1,offset)     = columns of matrix C beginning at offset
      mtype = 2
      do 60 i=1, NPROC
      call pvmfrecv(-1, mtype, rcode)
      call pvmfunpack(INTEGER4, offset, 1, 1, rcode)
      call pvmfunpack(INTEGER4, cols, 1, 1, rcode)
      call pvmfunpack(REAL8, c(1,offset), cols*NRA, 1, rcode)
  60  continue
```

```
C  Print results
      do 90 i=1, NRA
         do 80 j = 1, NCB
         write(*,70)c(i,j)
 70      format(2x,f8.2,$)
 80      continue
      print *, ' '
 90   continue

C  task now exits from PVM
      call pvmfexit(rcode)
      end
```

PARALLEL MATRIX MULTIPLY PROGRAM IN FORTRAN USING PVM FOR INTERPROCESSOR COMMUNICATION: WORKER PROGRAM

```
C*******************************************************************************
C PVM Matrix Multiply - Fortran Version
C Worker Program
C FILE: pvm_mm.worker.f
C DESCRIPTION: See pvm_mm.master.f
C PVM VERSION: 3.x
C LAST REVISED: 4/18/94 Blaise Barney
C*******************************************************************************
C Explanation of constants and variables used in this program:
C    NRA              = number of rows in matrix A
C    NCA              = number of columns in matrix A
C    NCB              = number of columns in matrix B
C    wtid             = PVM task id of this worker program
C    mtid             = PVM task id of master task
C    mtype            = PVM message type
C    cols             = columns of matrix B sent to each worker
C    offset        = starting position within the matrix
C    rcode, i, j, k   = misc.
C    a                = matrix A to be multiplied
C    b                = matrix B to be multiplied
C    c                = result matrix C
C - - - - - - - - - - - - - - - - - - - - - - - - - - - - - - - - - - - - - -
```

```
      program mm_worker
C     PVM Version 3.0 include file
      include 'fpvm3.h'

      parameter(NRA = 62)
      parameter(NCA = 15)
      parameter(NCB = 7)

      integer           wtid, mtid, mtype, cols, rcode, offset, i, j, k
      real*8    a(NRA,NCA), b(NCA,NCB), c(NRA,NCB)

C Enroll worker task
      call pvmfmytid(wtid)


C Receive message from master.  First set message type and determine the
C tid of the parent process.  Then receive following data from master:
C    offset = starting position in matrix
C    cols     = number of columns of B to receive
C    a      = receive all of matrix A
C    b        = receive some columns from matrix B
      mtype = 1
      call pvmfparent(mtid)
      call pvmfrecv(mtid, mtype, rcode)
      call pvmfunpack(INTEGER4, offset, 1, 1, rcode)
      call pvmfunpack(INTEGER4, cols, 1, 1, rcode)
      call pvmfunpack(REAL8, a, NRA*NCA, 1, rcode)
      call pvmfunpack(REAL8, b, cols*NCA, 1, rcode)

      write(*,10) wtid, cols
 10   format('worker task id = ',i8,' received ',i3,' cols from B')

C Do matrix multiply
         do 20 k=1, cols
           do 20 i=1, NRA
           c(i,k) = 0.0
             do 20 j=1, NCA
             c(i,k) = c(i,k) + a(i,j) *b(j,k)
 20          continue

C Set up send message to master. First set message type and
C initialize send buffer.  Then send following data elements to master:
C    offset = starting position in result matrix C
C    cols               = number of columns to send
C    c      = our part of result matrix C
      mtype = 2
```

```
      call pvmfinitsend(PVMDEFAULT, rcode)
      call pvmfpack(INTEGER4, offset, 1, 1, rcode)
      call pvmfpack(INTEGER4, cols, 1, 1, rcode)
      call pvmfpack(REAL8, c, cols*NRA, 1, rcode)

C  Send to master
      Call pvmfsend(mtid, mtype, rcode)

C  Exit PVM
      call pvmfexit(rcode)
      end


# # # # # # # # # # # # # # # # # # # # # # # # # # # # # # # # # # # # # # #
# PVM Matrix Multiply Makefile
# FILE make.pvm_mm.f
# DESCRIPTION: See pvm_mm.master.f
# PVM VERSION: 3.x
# USE: make -f make.pvm_mm.f
# LAST REVISED: 4/18/94 Blaise Barney
# # # # # # # # # # # # # # # # # # # # # # # # # # # # # # # # # # # # # # #


F77      =      xlf
PVMDIR   =      ${WORKSHOP}/pvm3
INCLUDE  =      -I${PVMDIR}/include
LIBS     =      -L${PVMDIR}/lib/RS6K -lfpvm3 -lpvm3


mm:     mm.master mm.worker


mm.master:  pvm_mm.master.f
    ${F77} pvm_mm.master.f ${INCLUDE} ${LIBS} -o mm.master


mm.worker:  pvm_mm.worker.f
    ${F77} pvm_mm.worker.f ${INCLUDE} ${LIBS} -o mm.worker
```

CONCURRENT WAVE EQUATION [3]

Description: This program implements the concurrent wave equation described in reference [1].

A vibrating string is decomposed into points. In the parallel version, each processor is responsible for updating the amplitude of a number of points over time. At each iteration, each processor exchanges boundary points with their nearest neighbors.

An X-based display of the final wave is provided for the parallel versions of
this code.

SERIAL VERSION OF WAVE EQUATION
PROGRAM IN FORTRAN

```
C   - - - - - - - - - - - - - - - - - - - - - - - - - - - - - - - - - - -
C   Serial Example - Wave Equation - Fortran Version
C   FILE: ser_wave.f
C   OTHER FILES: make.wave.f
C   DESCRIPTION:
C     This program implements the concurrent wave equation described
C     in Chapter 5 of Fox et al., 1988, Solving Problems on Concurrent
C     Processors, vol 1.
C
C     A vibrating string is decomposed into points.  In the parallel
C     version, each processor is responsible for updating the amplitude
C     of a number of points over time.
C
C     At each iteration, each processor exchanges boundary points with
C     nearest neighbors.
C
C   AUTHOR: R. Arnowitz
C   LAST REVISED: 4/18/94 Blaise Barney
C   - - - - - - - - - - - - - - - - - - - - - - - - - - - - - - - - - - -
C   Explanation of constants and variables used in common blocks and
C   include files
C     tpoints          = total points along wave
C     nsteps           = number of time steps
C     values(0:1001)   = values at time t
C     oldval(0:1001)   = values at time (t-dt)
C     newval(0:1001)   = values at time (t+dt)
C   - - - - - - - - - - - - - - - - - - - - - - - - - - - - - - - - - - -

      program wave
      implicit none


      print *, 'Serial Wave Program Running'


C     Get program parameters and initialize wave values
      call init_param
      call init_line
```

```
C     Update values along the wave for nstep time steps
          call update

      end


C     - - - - - - - - - - - - - - - - - - - - - - - - - - - - - - - - - - - -
C     Obtains input values from user
C     - - - - - - - - - - - - - - - - - - - - - - - - - - - - - - - - - - - -

      subroutine init_param

      implicit none
      integer tpoints, nsteps
      common/inputs/tpoints, nsteps

      integer MAXPOINTS, MAXSTEPS
      integer MINPOINTS
      parameter (MAXPOINTS = 1000)
      parameter (MAXSTEPS = 10000)
      parameter (MINPOINTS = 20)

      tpoints = 0
      nsteps = 0

      do while ((tpoints .lt. MINPOINTS) .or. (tpoints .gt. MAXPOINTS))
          write (*,*)'Enter number of points along vibrating string'
          read (*,*) tpoints
          if ((tpoints .lt. MINPOINTS) .or. (tpoints .gt. MAXPOINTS))
     &    write (*,*) 'enter value between ',MINPOINTS,' and ',MAXPOINTS

      end do

      do while ((nsteps .lt. 1) .or. (nsteps .gt. MAXSTEPS))
          write (*,*) 'Enter number of time steps'
          read (*,*) nsteps
          if ((nsteps .lt. 1) .or. (nsteps .gt. MAXSTEPS))
     &          write (*,*) 'enter value between 1 and ', MAXSTEPS
      end do

      write (*,10) tpoints, nsteps
 10   format(' points = ', I5, ' steps = ', I5)

      end


C     - - - - - - - - - - - - - - - - - - - - - - - - - - - - - - - - - - - -
C     Initialize points on line
C     - - - - - - - - - - - - - - - - - - - - - - - - - - - - - - - - - - - -
```

```
      subroutine init_line

      implicit none


      integer tpoints, nsteps
      common/inputs/tpoints, nsteps

      integer npoints, first
      common/decomp/npoints, first

      real*8 values(0:1001), oldval(0:1001), newval(0:1001)
      common/data/values, oldval, newval

      real*8 PI
      parameter (PI = 3.14159265)
      integer i, j, k
      real*8 x, fac

C     Calculate initial values based on sine curve
      fac = 2.0 * PI

      k = 0
      do j = 1, tpoints
         x = float(k)/float(tpoints - 1)
         values(j) = sin (fac * x)
         k = k + 1
      end do
      do i = 1, tpoints
         oldval(i) = values(i)
      end do
      end


C     - - - - - - - - - - - - - - - - - - - - - - - - - - - - - - - - -
C     Calculate new values using wave equation
C     - - - - - - - - - - - - - - - - - - - - - - - - - - - - - - - - -

      subroutine do_math(i)
      implicit none
      integer i

      integer tpoints, nsteps
      common/inputs/tpoints, nsteps

      real*8 values(0:1001), oldval(0:1001), newval(0:1001)
      common/data/values, oldval, newval
```

```
      real*8 dtime, c, dx, tau, sqtau

      dtime = 0.3
      c = 1.0
      dx = 1.0
      tau = (c * dtime / dx)
      sqtau = tau * tau
      newval(i) = (2.0 * values(i)) - oldval(i)
     &     + (sqtau * (values(i-1) - (2.0 * values(i)) + values(i+1)))
      end



C    - - - - - - - - - - - - - - - - - - - - - - - - - - - - - - - - - - -
C    Update all values along line a specified number of times
C    - - - - - - - - - - - - - - - - - - - - - - - - - - - - - - - - - - -

      subroutine update
      implicit none

      integer tpoints, nsteps, tpts
      common/inputs/tpoints, nsteps

      real*8 values(0:1001), oldval(0:1001), newval(0:1001)
      common/data/values, oldval, newval

      integer i, j

C    Update values for each point along string
      do i = 1, nsteps

C        Update points along line
         do j = 1, tpoints
C           Global endpoints
            if ((j .eq. 1).or.(j .eq. tpoints))then
               newval(j) = 0.0
            else
               call do_math(j)
            end if
         end do

         do j = 1, tpoints
            oldval(j) = values(j)
            values(j) = newval(j)
         end do

      end do
```

```
     if (tpoints .lt. 10) then
        tpts = tpoints
     else
        tpts = 10
     end if
     write (*,200) tpts, (values(i), i = 1, tpts)
200  format('first ', I5, ' points (for validation):'/
    &       10(f4.2, ' '))
     end
```

```
# # # # # # # # # # # # # # # # # # # # # # # # # # # # # # # # # # # # # # # # # #
# Serial wave Example Makefile
# FILE: make.ser_wave.f
# DESCRIPTION:  See ser_wave.f
# USE: make -f make.ser_wave.f
# LAST REVISED: 4/18/94 Blaise Barney
# # # # # # # # # # # # # # # # # # # # # # # # # # # # # # # # # # # # # # # # # #
F77        =        xlf

wave:      ser_wave.f
    ${F77} ser_wave.f -o wave
```

PARALLEL VERSION OF WAVE EQUATION PROGRAM IN FORTRAN USING PVM FOR INTERPROCESSOR COMMUNICATION

```
C  - - - - - - - - - - - - - - - - - - - - - - - - - - - - - - - - - - - - -
C  FILE: pvm_wave.f
C  OTHER FILES: make.pvm_wave.f, pvm_wave.h, draw_wave.c
C  DESCRIPTION:
C    This program implements the concurrent wave equation described
C    in Chapter 5 of Fox et al., 1988, Solving Problems on Concurrent
C    Processors, vol 1.
C
C    A vibrating string is decomposed into points.  Each processor is
C    responsible for updating the amplitude of a number of points over
C    time.  At each iteration, each processor exchanges boundary points with
C    nearest neighbors.
C
C  AUTHOR: Blaise Barney - adapted from MPL version by Roslyn Leibensperger
C  REVISED: 4/21/94 Blaise Barney
```

```
C   - - - - - - - - - - - - - - - - - - - - - - - - - - - - - - - - - - - - - -
C   - - - - - - - - - - - - - - - - - - - - - - - - - - - - - - - - - - - - - -
C   Explanation of constants and variables
C     NPROC           = number of tasks - must be at least 2 for this example
C     MAXPOINTS       = maximum number of points
C     MAXSTEPS        = maximum number of steps
C     BEGIN           = message type
C     DONE            = message type
C     RtoL            = message type
C     LtoR            = message type
C     NONE            = indicates no neighbor condition
C     DONTCARE        = accept a message from any task
C     PI              = value of pi
C     taskid          = task ID
C     parent          = parent task ID
C     tids()          = array to hold task ids
C     narch           = number of architectures
C     nsteps          = number of time steps
C     tpoints         = total points along wave
C     npoints         = number of points handled by this task
C     first           = index of first point handled by this task
C     rc              = generic return code
C     values()        = values at time t
C     oldval()        = values at time (t-dt)
C     newval()        = values at time (t+dt)
C   - - - - - - - - - - - - - - - - - - - - - - - - - - - - - - - - - - - - - -

      program wave
C     PVM version 3 include file
      include 'fpvm3.h'
      include 'pvm_wave.h'

C     Routine for creating the X graph of the wave
      external draw_wave

      integer i, left, right
C     Enroll process and find parent task id
      call pvmfmytid(taskid)
      call pvmfparent(parent)

C     *********************** parent code ***************************
      if (parent .eq. PvmNoParent) then
        print *, 'parent taskid=',taskid
        tids(1) = taskid
        do i = 2, NPROC
          call pvmfspawn("wave", PVMDEFAULT, " ", 1, tids(i), rc)
          print *, '...spawned child taskid=',tids(i)
```

```
      end do

      call get_input

C     send total points, time steps and tids list
      call pvmfinitsend(PVMDEFAULT,rc)
      call pvmfpack(INTEGER4,tpoints,1,1,rc)
      call pvmfpack(INTEGER4,nsteps,1,1,rc)
      call pvmfpack(INTEGER4, tids,NPROC,1,rc)
      do i = 2, NPROC
        call pvmfsend(tids(i),BEGIN,rc)
      end do
    end if


C     *********************** child code ***************************
    if (parent .ne. PvmNoParent)  then
C     receive total points, time steps and tids list
      call pvmfrecv(parent,BEGIN,rc)
      call pvmfunpack(INTEGER4,tpoints,1,1,rc)
      call pvmfunpack(INTEGER4,nsteps,1,1,rc)
      call pvmfunpack(INTEGER4,tids,NPROC,1,rc)
    end if


C     *********************** common code ***************************
C     determine left and right neighbors
      do i = 1, NPROC
        if (taskid .eq. tids(i)) then
          left = i-1
          right = i+1
        end if
      end do

      if (left .lt. 1) then
        left = NONE
      end if
      if (right .gt. NPROC) then
        right = NONE
      end if

      call init_line
C     Update values along the wave for nstep time steps
      call update(left, right)

C     parent collects results from child and prints
      if (taskid .eq. tids(1)) then
```

```fortran
          call output_parent
       else
          call output_child
       end if

C     Exit PVM
       call pvmfexit(rc)
       end

C   - - - - - - - - - - - - - - - - - - - - - - - - - - - - - - - - - - -
C     Gets input values from user - called by parent task
C   - - - - - - - - - - - - - - - - - - - - - - - - - - - - - - - - - - -

       subroutine get_input

       include 'fpvm3.h'
       include 'pvm_wave.h'

       tpoints = 0
       nsteps = 0

       do while ((tpoints .lt. NPROC) .or. (tpoints .gt. MAXPOINTS))
          write (*,*)'Enter number of points along vibrating string'
          read (*,*) tpoints
          if ((tpoints .lt. NPROC) .or. (tpoints .gt. MAXPOINTS))
     &        write (*,*) 'enter value between ',NPROC,' and ',MAXPOINTS
       end do

       do while ((nsteps .lt. 1) .or. (nsteps .gt. MAXSTEPS))
          write (*,*) 'Enter number of time steps'
          read (*,*) nsteps
          if ((nsteps .lt. 1) .or. (nsteps .gt. MAXSTEPS))
     &           write (*,*) 'enter value between 1 and ' MAXSTEPS
       end do

       write (*,10) taskid, tpoints, nsteps
10     format(I8, ': points = ', I5, ' steps = ', I5)

       end


C   - - - - - - - - - - - - - - - - - - - - - - - - - - - - - - - - - - -
       Initialize points on line - each task does its portion
C   - - - - - - - - - - - - - - - - - - - - - - - - - - - - - - - - - - -

       subroutine init_line
       include 'fpvm3.h'
```

```
      include 'pvm_wave.h'

      integer nmin, nleft, npts, i, j, k
      real*8 x, fac

C     Determine even distribution of points and number leftover
      nmin = tpoints/NPROC
      nleft = mod(tpoints, NPROC)

C     Calculate initial values based on sine curve
      fac = 2.0 * PI
      k = 0
      do i = 1, NPROC
         if (i .lt. nleft) then
            npts = nmin + 1
         else
            npts = nmin
         end if
         if (taskid .eq. tids(i)) then
            first = k + 1
            npoints = npts
            write (*,15) taskid, first, npts
15          format (I8, ': first = ', I5, ' npoints = ', I5)
            do j = 1, npts
               x = float(k)/float(tpoints - 1)
               values(j) = sin (fac * x)
               k = k + 1
            end do
         else
            k = k + npts
         end if
      end do
      do i = 1, npoints
         oldval(i) = values(i)
      end do
      end


C     - - - - - - - - - - - - - - - - - - - - - - - - - - - - - - - -
C        All processes update their points a specified number of times
C     - - - - - - - - - - - - - - - - - - - - - - - - - - - - - - - -

      subroutine update(left, right)

      include 'fpvm3.h'
      include 'pvm_wave.h'

      integer left, right, i, j
```

```
        real*8 dtime, c, dx, tau, sqtau

        dtime = 0.3
        c = 1.0
        dx = 1.0
        tau = (c *dtime / dx)
        sqtau = tau * tau


C       Update values for each point along string
        do i = 1, nsteps
C           Exchange data with "left-hand" neighbor - if you have one
            if (left .ne. NONE) then
              call pvmfinitsend(PVMDEFAULT,rc)
              call pvmfpack(REAL8,values(1),1,1,rc)
              call pvmfsend(tids(left),RtoL,rc)

              call pvmfrecv(tids(left),LtoR,rc)
              call pvmfunpack(REAL8,values(0),1,1,rc)
            end if

C           Exchange data with "right-hand" neighbor
            if (right .ne. NONE) then
              call pvmfinitsend(PVMDEFAULT,rc)
              call pvmfpack(REAL8,values(npoints),1,1,rc)
              call pvmfsend(tids(right),LtoR,rc)

              call pvmfrecv (tids(right),RtoL,rc)
              call pvmfunpack(REAL8,values(npoints+1),1,1,rc)
            end if

C           Update points along line
            do j = 1, npoints
C             Global endpoints
              if ((first+j-1 .eq. 1).or.(first+j-1 .eq. tpoints)) then
                newval(j) = 0.0
              else
C               Use wave equation to update points
                newval(j) = (2.0 * values(j)) - oldval(j)
     &          + (sqtau * (values(j-1) - (2.0 * values(j)) + values(j+1)))
              end if
            end do

            do j = 1, npoints
              oldval(j) = values(j)
              values(j) = newval(j)
            end do
```

```
         end do
         print *, 'task-',taskid,'done with update'
         end

C  - - - - - - - - - - - - - - - - - - - - - - - - - - - - - - - - - -
C      Parent receives results from child and prints and shows graph
C  - - - - - - - - - - - - - - - - - - - - - - - - - - - - - - - - - -
         subroutine output_parent

         include 'fpvm3.h'
         include 'pvm_wave.h'

         integer i, start, npts, tpts
         real*8 results(MAXPOINTS)

C      Store child's results in results array
         do i = 2, NPROC
C        Receive number of points, first point and results
            call pvmfrecv(DONTCARE,DONE,rc)
            call pvmfunpack(INTEGER4,start,1,1,rc)
            call pvmfunpack(INTEGER4,npts,1,1,rc)
            call pvmfunpack(REAL8,results(start),npts,1,rc)
         end do

C      Store parent's results in results array
         do i = first, first+npoints-1
            results(i) = values(i)
         end do

         if (tpoints .lt. 10) then
            tpts = tpoints
         else
            tpts = 10
         end if
         write (*,200) tpts, (results(i), i = 1, tpts)

C      Display results with draw_wave routine
         call draw_wave(%REF(results))

 200     format('first ', I5, ' points (for validation):'/
        &        10(f4.2, '  '))
         end

C  - - - - - - - - - - - - - - - - - - - - - - - - - - - - - - - - - -
C      Send the updated values to the parent
C  - - - - - - - - - - - - - - - - - - - - - - - - - - - - - - - - - -
```

```
      subroutine output_child

      include 'fpvm3.h'
      include 'pvm_wave.h'

C     send first point, number of points and results to parent
      call pvmfinitsend(PVMDEFAULT,rc)
      call pvmfpack(INTEGER4,first,1,1,rc)
      call pvmfpack(INTEGER4,npoints,1,1,rc)
      call pvmfpack(REAL8,values(1),npoints,1,rc)
      call pvmfsend(parent,DONE,rc)
      end
```

```
# # # # # # # # # # # # # # # # # # # # # # # # # # # # # # # # # # # # # # #
# FILE: make.pvm_wave.f
# DESCRIPTION:  see pvm_wave.f
# USE: make -f make.pvm_wave.f
# LAST REVISED: 04/21/94 Blaise Barney
# # # # # # # # # # # # # # # # # # # # # # # # # # # # # # # # # # # # # # #

F77     =    xlf
CC      =    cc
OBJ     =    wave
SRC     =    pvm_wave.f draw_wave.o
PVMDIR  =    ${WORKSHOP}/pvm3
INCLUDE =    -I${PVMDIR}/include
LIBS    =    -L${PVMDIR}/lib/RS6K -lfpvm3 -lpvm3  -lm
XLIBS   =    /usr/lib/libX11.a

${OBJ}: ${SRC}
      ${F77} ${SRC} ${INCLUDE} ${LIBS} ${XLIBS} -o ${OBJ}

draw_wave.o: draw_wave.c
      ${CC} -c draw_wave.c
C  - - - - - - - - - - - - - - - - - - - - - - - - - - - - - - - - - - - - -
C     Fortran include file for pvm_wave.f
C  - - - - - - - - - - - - - - - - - - - - - - - - - - - - - - - - - - - - -
      implicit none

      integer BEGIN,DONE,RtoL,LtoR,NPROC,MAXPOINTS,MAXSTEPS,
     &        NONE,DONTCARE
      parameter (BEGIN = 1)
      parameter (DONE = 2)
      parameter (RtoL = 3)
      parameter (LtoR = 4)
      parameter (NPROC = 4)
      parameter (MAXPOINTS = 1000)
```

```
        parameter (MAXSTEPS = 10000)
        parameter (NONE = -1)
        parameter (DONTCARE = -1)

        real*8 PI
        parameter (PI = 3.14159265)


        integer taskid, parent, tids(NPROC), narch, rc
        common/pvmvars/taskid, parent, tids, narch, rc

        integer tpoints, nsteps
        common/inputs/tpoints, nsteps

        integer npoints, first
        common/decomp/npoints, first

        real*8 values(0:MAXPOINTS+1), oldval(0:MAXPOINTS+1),
     &        newval(0:MAXPOINTS+1)
        common/data/values, oldval, newval
* - - - - - - - - - - - - - - - - - - - - - - - - - - - - - - - - - -
* FILENAME: draw_wave.c
* OTHER FILES:  see pvm_wave.c or pvm_wave.f
* DESCRIPTION:  Called by pvm_wave to draw a graph of results
* AUTHOR: Blaise Barney
* LAST REVISED: 4/21/94 Blaise Barney
* - - - - - - - - - - - - - - - - - - - - - - - - - - - - - - - - - -

#include <stdio.h>
#include <Xll/Xlib.h>
#include <Xll/Xutil.h>

#define HEIGHT          500
#define WIDTH           1000

typedef struct {
   Window window;
   XSizeHints hints;
   XColor backcolor;
   XColor bordcolor;
   int    bordwidth;
} MYWINDOW;

char baseword[] = {"draw_wave"},
     exitword[] = {"Exit"},
     text[10];


void draw_wave(double * results) {
```

```
float         scale, point, coloratio = 65535.0 / 255.0;
int    i,j,k,y, zeroaxis, done, myscreen, points[WIDTH];

MYWINDOW base, quit;
Font font,font2;
GC            itemgc,textgc,pointgc,linegc;
XColor        red,yellow,blue,green,black,white;
XEvent        myevent;
Colormap cmap;
KeySym        mykey;
Display *mydisp;


/* Set rgb values for colors */
red.red= (int) (255 * coloratio);
red.green= (int) (0 * coloratio);
red.blue = (int) (0 * coloratio);

yellow.red= (int) (255 * coloratio);
yellow.green= (int) (255 * coloratio);
yellow.blue= (int) (0 * coloratio);

blue.red= (int) (0 * coloratio);
blue.green= (int) (0 * coloratio);
blue.blue= (int) (255 * coloratio);

green.red= (int) (0 * coloratio);
green.green= (int) (255 * coloratio);
green.blue= (int) (0 * coloratio);

black.red= (int) (0 * coloratio);
black.green= (int) (0 * coloratio);
black.blue= (int) (0 * coloratio);

white.red= (int) (255 * coloratio);
white.green= (int) (255 * coloratio);
white.blue= (int) (255 * coloratio);

mydisp = XOpenDisplay("");
if (!mydisp) {
   fprintf (stderr, "Hey! Either you don't have X or something's not
right.\n");
   fprint (stderr, "Guess I won't be showing the graph.  No big deal.\n");
   exit(1);
   }
myscreen = DefaultScreen(mydisp);
```

```
cmap = DefaultColormap (mydisp, myscreen);
XAllocColor (mydisp, cmap, &red);
XAllocColor (mydisp, cmap, &yellow);
XAllocColor (mydisp, cmap, &blue);
XAllocColor (mydisp, cmap, &black);
XAllocColor (mydisp, cmap, &green);
XAllocColor (mydisp, cmap, &white);


/* Set up for creating the windows */
/* XCreateSimpleWindow uses defaults for many attributes,    */
/* thereby simplifying the programmer's work in many cases. */

/* base window position and size */
base.hints.x = 50;
base.hints.y = 50;
base.hints.width = WIDTH;
base.hints.height = HEIGHT;
base.hints.flags = PPosition | PSize;
base.bordwidth = 5;


/* window Creation */
/* base window */
base.window = XCreateSimpleWindow (mydisp, DefaultRootWindow (mydisp),
                base.hints.x, base.hints.y, base.hints.width,
                base.hints.height, base.bordwidth, black.pixel,
                black.pixel);
XSetStandardProperties (mydisp, base.window, baseword, baseword, None,
                    NULL, 0, &base.hints);

/* quit window position and size (subwindow of base) */
   quit.hints.x = 5;
   quit.hints.y = 450;
   quit.hints.width = 70;
   quit.hints.height = 30;
   quit.hints.flags = PPosition | PSize;
   quit.bordwidth = 5;


   quit.window = XCreateSimpleWindow (mydisp, base.window, quit.hints.x,
                    quit.hints.y, quit.hints.width, quit.hints.height,
                    quit.bordwidth, green.pixel, yellow.pixel);
XSetStandardProperties (mydisp, quit.window, exitword, exitword, None,
            NULL, 0, &quit.hints);

/* Load fonts */
/*
font = XLoadFont (mydisp, "Rom28");
font2 = XLoadFont (mydisp, "Rom17.500");
```

```
*/
font = XLoadFont (mydisp, "fixed");
font2 = XLoadFont (mydisp, "fixed");

/* GC creation and initialization */
textgc = XCreateGC (mydisp, base.window, 0,0);
XSetFont (mydisp, textgc, font);
XSetForeground (mydisp, textgc, white.pixel);

Linegc = XCreateGC (mydisp, base.window, 0,0);
XSetForeground (mydisp, linegc, white.pixel);

itemgc = XCreateGC (mydisp, quit.window, 0,0);
XSetFont (mydisp, itemgc, font2);
XSetForeground (mydisp, itemgc, black.pixel);

pointgc = XCreateGC (mydisp, base.window, 0,0);
XSetForeground (mydisp, pointgc, green.pixel);

/* The program is event driven; the XSelectInput call sets which */
/* kinds of interrupts are desired for each window. */
/* These aren't all used. */
XSelectInput (mydisp, base.window,
              ButtonPressMask | KeyPressMask | ExposureMask);
XSelectInput (mydisp, quit.window,
              ButtonPressMask | KeyPressMask | ExposureMask);

/* window mapping - this lets windows be displayed */
XMapRaised (mydisp, base.window);
XMapSubwindows (mydisp, base.window);

/* Scale each data point */
zeroaxis = HEIGHT/2;
scale = (float)zeroaxis;
for(j=0;j<WIDTH;j++)
   points[j]  = zeroaxis - (int)(results[j] * scale);

/* Main event loop - exits when user clicks on "exit" */
done = 0;
while (! done) {
   XNextEvent (mydisp, &myevent);  /* Read next event */
   switch (myevent.type) {
   case Expose:
      if (myevent.xexpose.count == 0) {
         if (myevent.xexpose.window == base.window) {
            XDrawString (mydisp, base.window, textgc, 775, 30, "Wave",4);
            XDrawLine (mydisp, base.window, linegc, 1,zeroaxis,WIDTH,
```

```
                         zeroaxis);
               for (j=1; j<WIDTH; j++)
                   XDrawPoint (mydisp, base.window, pointgc, j, points[j-1]);
               }

            else if (myevent.xexpose.window == quit.window) {
                XDrawString (mydisp, quit.window, itemgc, 12,20, exitword,
                             strlen(exitword));
                }
          }   /* case Expose */
          break;

      case ButtonPress:
         if (myevent.xbutton.window == quit.window)
            done = 1;
         break;

      case KeyPress;
         i = XLookupString (&myevent, text, 10, &mykey 0);
         if (i == 1 && text[0] == 'q')
            done = 1;
         break;

      case MappingNotify:
         XRefreshKeyboardMapping (&myevent);
         break;

         } /* switch (myevent.type) */

      }  /* while (! done) */

XDestroyWindow (mydisp, base.window);
XCloseDisplay (mydisp);
exit (0);
}
```

References

1. Array Assignment Program, Blaise Barney, Maui High Performance Computing Center, Kihei, HI.
2. Matrix Multiply Program, Ros Leibensperger and Blaise Barney, Maui High Performance Computing Center, Kihei, HI.
3. Concurrent Wave Equation Program, R. Arnowitz and Blaise Barney, Maui High Performance Computing Center, Kihei, HI.

Glossary

Ada: A computer programming language sponsored by the US Dept. of Defense.

ADI: Alternative direction implicit; a numerical analysis method.

ALU: Arithmetic Logic Unit.

Array processor: A type of parallel computer designed to operate most efficiently on arrays of data such as digital images or fluid dynamics applications.

Associative processor: A type of parallel computer with associative, that is, content addressable memory.

Bandwidth: The rate at which information travels from place to place.

Barriers: A programming construct for MIMD computers that causes processes to wait and allows them to proceed only after a predetermined number of processes are waiting at the barrier.

Bisection bandwidth: The number of datapaths that cross the bisection (an imaginary surface that divides a parallel computer into two equal halves) times the data rate in each datapath.

Butterfly switch: A routing network employed by the Butterfly computer.

C: A popular programming language.

Cache: Temporary memory, usually relatively small, like a scratchpad.

Chrysalis: Operating system for the Butterfly.

Coterie network: In the Image Understanding Architecture, an interconnected processor subset temporarily segregated from other processors; from the French *coterie,* a social circle or clique.

CPU: Central Processing Unit.

DCE: Distributed Computing Environment (software).

Debugger: A program development tool that facilitates the isolation of programming errors.

Distributed memory: A parallel computer architectural characteristic in which parts of the total memory are associated with and local to each processor.

DMA channel: Direct Memory Access data path.

Do loop: A program segment that is executed repeatedly for a specified number of iterations or until a condition is satisfied.

DOE: US Dept of Energy.

DRAM: Dynamic read and write memory.

FFT: Fast Fourier transform.

FIFO: First in, first out scheduling strategy.

Floating point operation: Arithmetic operation (e.g., add or multiply) involving float point numbers (i.e., numbers with decimal points).

Fork: Create a parallel process.

Fortran: A high level programming language, usually for scientific or engineering applications.

Heterogeneous architecture: A multiprocessor in which the processing elements differ from each other.

HiPPI: High performance peripheral interface.

Host: A computer, usually a workstation, from which a parallel processor receives I/O services, a user interface, debugging and load balancing tools, diagnostics, and related ancillary functions.

Hypercube: An interconnect paradigm based on connecting the corners of a cube in N-dimensional space.

Join: Destroy a parallel process.

Kalman filter: A mathematical estimation method, used in tracking.

LAN: Local area network.

Latency: The time delay experienced before a process begins.

LISP: A programming language effective for list processing; used in artificial intelligence.

Load balance: The act of assigning processes to the processors of a parallel computer so that all of the processors have approximately the same amount of computing to perform.

Master/slave paradigm: A programming strategy for MIMD computers in which worker programs on processors designated "slaves" are controlled by the main program on a single other processor.

Memory lock: See spin lock; prevents processors from accessing protected memory until a condition is satisfied.

Memory shares: Partitions of the memory that can be accessed by all processors.

MIMD: Multiple instruction, multiple datastream.

MPI: Message passing interface.

MPP: Massively parallel processor.

MTA: Multithreaded architecture.

Multiprocessor: A computer with multiple processors; these may be programmed to act in parallel on a single program or programmed to perform independent, unrelated tasks.

Neural net: A computational method, usually implemented in hardware, in which a highly connected collection of very simple processing nodes each respond with an output signal conditioned by a weighted sum of input signals; the neural net is programmed by setting the weights through training on repetitive examples with known outputs.

NUMA: Nonuniform memory access.

OLTP: On-line transaction processing.

Operand: Part of a computer machine instruction.

Packet: Each part into which a long message is partitioned in a packet switched communications system.

Parallel processing: The use of multiple processors simultaneously working on one problem.

Parallelizing compiler: A software tool that converts a program in a high level language (e.g., C or Fortran) into machine instructions and that finds ways to disperse the computation over multiple processors for simultaneous execution.

Partition: (1) A subset of the processor set of a parallel computer devoted to one program; (2) A portion of memory reserved as private to one processor or designated as shared.

Pixel: A picture element; in digital images, the information is encoded as rows of pixels, with each pixel valued as the relative brightness of that point in the image.

Portability: A feature of some programs that allows them to be executed on diverse computers.

Process: A group of instructions that can be assigned for execution to a single processor.

Processing element: Each processor in a multiprocessor computer.

Processing ensemble: A collection of processors under control of a single control unit.

RISC: Reduced instruction set computer.

Router: A hardware device that performs message routing.

Routing: The choosing of an information path from among several.

SAR: Synthetic aperture radar.

Scalability: A parallel computer architectural characteristics in which performance increases in proportion with machine size.

Self-scheduling: A programming method for MIMD computers that allows a processor to take onto itself the next task in the queue whenever it is ready to do so.

Shared memory: A parallel computer architectural characteristic in which a single memory is monolithic and used by all of the processors in turn.

SIMD: Single instruction, multiple datastream.

SMP: Variously, Symmetric Multiprocessor or Shared-Memory Multiprocessor.

Spin lock: A programming construct for MIMD computers that prevents more than one processor from accessing protected memory simultaneously.

Supercomputer: That class of computers that share the features of high speed and large capacity compared with the average of what is available at any given time.

Synchronization: In SIMD computers, the property of having all processors execute instructions in lockstep; in MIMD computers, the coordination of multiple communicating processes to assure that a message between them is executed while they are in the expected state.

Task decomposition: The separation of a large program into smaller segments (tasks) that can be executed simultaneously with semi-independence.

TFLOP/s: Tera (a thousand billion) floating point operations per second.

Tightly coupled: A feature of some MIMD multiprocessors in which the number of processors devoted to a program remains fixed and they operate under the supervision of a strict control scheme.

Unix: A widely used operating system developed by AT&T.

Vector processor: A computer designed to be particularly effective on vector operations, for example, Cray C-90.

Very Large Scale Integration (VLSI): A measure of how many components are on a single computer chip.

Virtual processor: A software construct used for ease of programming in which a single physical processor is caused to behave like a number of processors.

Warp: A systolic array processor sponsored by DARPA (DoD).

Index

About the Author

R. Michael Hord is a national leader in high-performance computing and is presently a member of the senior technical staff at DBA Systems, a wholly owned subsidiary of the Titan Corporation. Until 1993, he was director of High Performance Computing at the General Electric Advanced Technology Laboratories (Moorestown, NJ). In this capacity, he directed research and development activities employing the Connection Machine, the Butterfly, and the Warp advanced architecture computers. The emphasis was on acoustic signal analysis, military data/information systems, and future architectures.

Until the end of 1989, Mr. Hord was head of the Advanced Development Center at MRJ, Inc. (Oakton, VA), where he directed diverse computer applications using parallel architectures including two Connection Machines. Application areas included image processing, signal processing, electromagnetic scattering, operations research, and artificial intelligence. Mr. Hord joined MRJ, where he also directed the corporate research and development program, in 1984.

For five years (1980–1984) Mr. Hord was the director of Space Systems for General Research Corporation. Under contract to NASA and the Air Force, he and his staff assessed technology readiness for future space systems and performed applications analysis for innovative on-board processor architectures.

SIMD parallel processing was the focus of his efforts as the manager of applications development for the Institute for Advanced Computation (IAC). IAC was the joint DARPA/NASA sponsored organization responsible for the development of the Illiac IV parallel supercomputer at Ames Research Center. Projects included computational fluid dynamics, seismic simulation, digital cartography, linear programming, climate modeling, and diverse image and signal processing applications.

Prior positions at Earth Satellite Corporation, Itek Corporation, and Technology Incorporated were devoted to the development of computationally intensive applications, such as optical system design and natural resource management.

Mr. Hord's seven prior books and scores of papers address advanced parallel computing, digital image processing, and space technology. He has long been active in the applied imagery pattern recognition community, has been an IEEE Distinguished Visitor, and is a guest lecturer at several universities. His B.S. in physics was granted by Notre Dame University in 1962 and in 1966 he earned an M.S. in physics from the University of Maryland.